TABLE OF CONTENTS

Experiment

N.B. Unless otherwise noted, photographs are by the author.

Parasitology-Related Websites and Resources

Professional Societies:
American Society of Parasitologists
The American Society of Parasitologists
P.O. Box 1897, Lawrence, KS 66044, U.S.A.
Tel: 1-800-627-0326 (or 1-785-865-9404)   Fax: 1-785-843-6153
World Wide Web Site:
        http://www-museum.unl.edu/asp/
Journal: *Journal of Parasitology*

American Society of Tropical Medicine and Hygiene
60 Revere Drive, Suite 500
Northbrook, Ill. 60062
(847) 480-9592
Fax: (847) 480-9283
E-Mail: astmh@aol.com
Journal: *American Journal of Tropical Medicine and Hygiene*

Useful Internet Sites:

Batch of Bug Sites
http://www.microbes.info

Merck Veterinary Manual
http://www.merckvetmanual.com

Home page for the journal *Molecular and Biochemical Parasitology*
http://www.elsevier.nl/cas/estoc/contents/SA1/01666851.html

World Health Organization Web Page (Contains WHO documents on tropical health)
http://www/who.ch

Directory of Parasitologists: Lists scientists working in the field
URL <ftp://magnus.acs.ohio-state.edu/pub/zoology>

Parasite Genome Projects: Provides descriptions of projects and links to other genome projects
http://woodland.bio.ic.ac.uk/fgn/parasite-genome/parasite-

genome.html

USENET groups:
Newsgroups for parasitology-related areas, including bionet.parasitology, bionet.molbio, and bionet.protista
Visit http://www.bio.net/ for all bionet postings

National Library of Medicine
http://www.nih.nlm.gov

PubMed
http://www.PubMed.gov

Ohio State University collection of Parasitology Images
http://www.biosci.ohio-state.edu/~parasite/

Center for Disease Control, Atlanta, GA
http://www.dpd.cdc.gov/DPDx

A fascinating account regarding epidemiology
http://www.ph.ucla.edu/epi/snow.html

Articles You May Not Want To Read Before Lunch

Stone, Richard. 2001. Down to the Wire on Bioweapons Talks. Science 293:414-416.

Colwell, Strant T., Jr. 1998. Prevalence of Helminths in Fecal Deposits of Dogs in Anderson County, Tennessee. Journal of the Tennessee Academy of Science 73(3-4):104-105.

Tongue-eating bug found in fish
http://news.bbc.co.uk/cbbcnews/hi/newsid_4200000/newsid_4209000/4209004.stm

SOME BASIC CONCEPTS
A Picture Painted with a Broad Brush
Parasitic infections are relatively rare in the United States. Why? Because most Americans:

1. Can afford shoes;
2. Have adequate nutrition, at least relative to calories and protein;
3. Have access to a water supply that is not contaminated with raw sewage;
4. Have adequate access to health care resources (medical professionals, nearby hospitals, antibiotics, drugs, vaccines);
5. Use synthetic fertilizers to grow crops, as opposed to human nightsoil;
6. Life in the temperate zone, where there is a season during which insect vectors are absent.

So? **People live long enough to show diseases of degeneration, such as:**

- Cancer: ~ 400,000 deaths per year
- Heart disease: ~ 800,000 deaths per year

This is NOT the case in 3rd World Countries:

World Life Expectancy
http://www.worldlifeexpectancy.com/

TABLE 1. ESTIMATES OF THE NUMBERS OF HELMINTH INFECTIONS IN MAN.
(ADAPTED FROM PETERS AND GILLES, 1977; PETERS, 1978 and Schmidt & Roberts, Foundations of Parasitology, ed. 5, 1996)

INTESTINAL NEMATODES	MILLIONS	Deaths per Year
All helminths	*3500*	
Ascaris lumbricoides	1250	20,000
Hookworm (*Necator* sp., *Ancylostoma* spp.)	950	50,000-60,000
Trichuris trichiura	700	
Enterobius vermicularis	350	
Strongyloides stercoralis	60	
TISSUE NEMATODES		
Wuchereria bancrofti	350	

Dracunculus medinensis	80	
Trichinella spiralis	50	
Onchocerca volvulus	40	
Loa loa	20	

TREMATODES
Schistosoma spp.	300	500 K to 1 million
Clonorchis sinensis	40	
Fasciolopsis buski	15	
Paragonimus westermanni	5	

CESTODES
Taenia spp.	80
Hymenolepis Spp.	40
Diphyllobothrium latum	15

PROTOZOA
(From Markell & Voge: Medical Parasitology)
Entamoeba histolytica	600	
Plasmodium Spp.	489	1-2 million
African trypanosomiasis	35	
American trypanosomiasis	10	

In 1986, there were an estimated 60 million deaths, of which 30 million are children <5 years old. Half of the deaths among children (15 million) were due to a combination of malnutrition and intestinal infection.

DEFINITIONS

<u>PARASITE</u>: An organism which derives sustenance or benefit at the expense of its host.

      <u>ENDOPARASITE</u> = internal

      <u>ECTOPARASITE</u> = external

    OBLIGATE: Parasite stage necessary for completion of life cycle.
        e.g., *Trichuris trichiura, Entamoeba histolytica.*

    FACULTATIVE: Normally free-living, but can exist as a parasite.
        e.g., *Strongyloides stercoralis.*

    ACCIDENTAL: A parasite found in an abnormal host.
        e.g., *Naegleria fowleri.*

<u>HOST</u>: That organism which is necessary for the development of a parasite.

    DEFINITIVE: Parasite reaches sexual maturity.
        e.g., Humans for *Clonorchis sinensis*

    INTERMEDIATE: Necessary for development, but parasite does not reach sexual maturity.
        e.g., cyprinid fishes for *Clonorchis sinensis*

    PARATENIC: A host which is not necessary for the physiological development of the parasite, but which facilitates transferral from the intermediate host to the definitive host to the definitive host.

    VECTOR: Any agent, e.g. insect, that transmits a disease organism.

<u>HYPERPARASITISM</u>: Parasite serving as a host for another parasitic species.

For Protozoa:
<u>TROPHOZOITE</u>: Metabolically active form of protozoan parasites,
    within the apppropriate organ of the host.

<u>CYST</u>: Metabolically inactive form of protozoan parasites,
    adapted for transmission.

ADAPTATIONS TO PARASITIC EXISTENCE

I. SPECIALIZATION

A. Certain sensory organs are highly developed,
   1. Chemosensory organs among cercariae for the detection of snail mucus.
   2. Thermosensory organs among hookworm larvae for the detection of homeothermic animals.

B. Resistant stage (e.g. cyst or egg) for transferral to new hosts.
   1. Cyst form of intestinal protists and helminth ova are metabolically inactive, but can survive reduced temperatures and increased osmotic pressures of freshwater.

II. DEGENERATION

A. "Unnecessary" organs become vestigial or disappear.
   1. Most internal helminths lack pigments and eyes.
   2. Cestodes do not have any digestive system -- nutrients are absorbed directly through the tegument.
B. Loss of capacity to synthesize enzymes, nutrients
   1. Trematodes lack the ability to synthesize certain fatty acids which must be in their diet, and this "flaw" lends itself to a commensal or parasitic existence.

III. HIGH BIOTIC POTENTIAL, facilitated by

A. Hermaphroditism: Adults have functional male and female reproductive systems. In parasitic organisms, this hermaphroditism is simultaneous, as opposed to sequential.
1. Characteristic of Platyhelminthes, including cestodes (tapeworms) and trematodes (flukes)
2. Cross-fertilization occurs across proglottids of adjacent cestodes or even the same cestode, and across adjacent flukes.

B. Parthenogenesis: Development of an egg without fertilization by a sperm cell. Common among insects, e.g. aphids.

C. Polyembryony: Larval forms undergo a form of budding in the intermediate host. This is observed in digenetic trematodes (flukes) and some wasps.

C. Strobilization: Segments (proglottids) develop behind a holdfast organ (scolex). Each proglottid develops within it complete male and female reproductive systems. Following self- or cross-fertilization, the male reproductive structures deteriorate and eggs mature mature. Proglottids which are furthest away from the scolex are gravid, and will release eggs when the proglottid is shed.

Adult *Echinococcus granulosus*, the small dog tapeworm, showing the anterior scolex with adhesive suckers, followed by a strobila consisting of one immature, one mature, and one gravid proglottid.

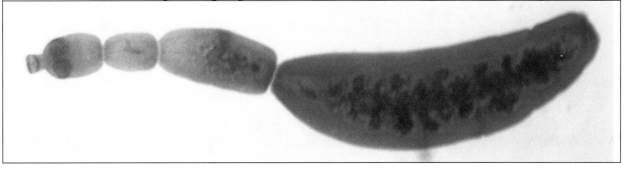

SIX ESSENTIAL ASPECTS TO A PARASITE LIFE CYCLE

1. Find a Host

    A. Active: Host produces either chemical, thermal or light signals to which the infective stage is sensitive.

    B. Passive: Infective stages are dispersed passively through environment, such as waves or water currents.

2. Enter a Host

    A. Active: Infective stage may burrow into skin, as cercariae of blood flukes do.

    B. Passive: Host will ingest or inhale infective forms.

3. Overcome Host Defenses: Mechanisms include

    A. Antigen Shielding: Surface of parasite adsorbs host derived antigen, so that parasite is recognized as "self". (Documented among adult blood flukes)

    B. Surface Antigen Shifting: Proteins forming protein surface change so that immune reactions lag behind development of the parasites. (Documented in *Trypanosoma gambiense* and *T. rhodesiense* (African trypanosomiasis.)

4. Derive Nutrients From Host

    A. Aerobic Metabolism

    B. Obligate Anaerobic Metabolism

5. Reproduce More Individuals

    A. Hermaphroditism

    B. Polyembrony

C. Very high egg output: among some tapeworms, daily output can be in millions, but parental care is negligible.

6. Disperse Young to New Hosts

   A. Presence of Obligate Free-living Stage: Needs to tolerate changes in temperature, osmotic pressure, desiccation.
   B. Passive or Active Dispersal: Cercariae of digenetic trematodes burrow out of snail host in search of 2nd intermediate host or of definitive host.
   C. Parasite-Induced Change in Host Behavior: Intermediate stages will induce qualitative changes in the intermediate host such that the infected intermediate host is more likely to be captured by a predator which is the definitive host.
   D. Parasite-induced change in host morphology, making them less able to avoid predators, as in the case of *Ribeiroia ondotrae* metacercariae among green frogs.

DIAGNOSTIC METHODS

I.   FECAL EXAMINATION   Intestinal Helminths
Intestinal Protozoa

II.   URINALYSIS   *Schistosoma haematobium*

III.   BLOOD SMEAR   *Plasmodium* sp. (Malaria)

IV.   IMMUNOASSAY   *Pneumocystis carinii*
Cysticercosis (*Taenia solium*)
Echinococcus granulosus Cysts
Giardia lamblia

V.   OCULAR EXAMINATION   *Toxoplasma gondii*
Toxocara spp.

VI.   BIOPSY   *Onchocerca volvulus*
Trichinella spiralis

VII. VAGINAL SMEAR   *Trichomonas vaginalis*

VIII. IMMUNOBLOT   *Plasmodium falciparum*

IX. PCR   Malaria (see *J Infect Dis 2009;199:1561-1563,1567-1574.*)

X.   XENODIAGNOSIS   *Trichinella spiralis*

CANDIDATES FOR PARASITIC INFECTIONS

I. IMMIGRANTS

 MALARIA, AMOEBIASIS, SCHISTOSOMIASIS, CLONORCHIASIS

II. TOURISTS & SERVICEMEN (NOT NECESSARILY OVERSEAS)

 SAME AS ABOVE PLUS GIARDIASIS

III. CHILDREN

 Enterobius (PINWORM), HYDATID CYST (*Echinococcus*), ASCARIASIS, TRICHURIASIS, HOOKWORMS, LICE, TOXOPLASMOSIS, DERMAL LARVA MIGRANS, VISCERAL LARVA MIGRANS, *Naegleria/Acanthomoeba*

IV. IMMUNOSUPPRESSED

 Pneumocystis, Toxoplasma, Strongyloides, Cryptosporidium

V. RURAL AND INDIGENT

 Trichinella, Strongyloides, Giardia, Entamoeba, HOOKWORMS, ASCARIASIS

VI. PEOPLE WHOSE DIETS INCLUDE RAW MEATS

 Trichinella (pork), *Diphyllobothrium* (freshwater fish), *Anisakis* (marine fish), *Taenia* (pork), *Taeniarhynchus* (beef)

From: http://www.Dribbleglass.com/subpages/billboards57b.htm

VIII. PROMISCUOUS (though not necessarily)

Trichomonas vaginalis

Phylogenetic Tree of Life

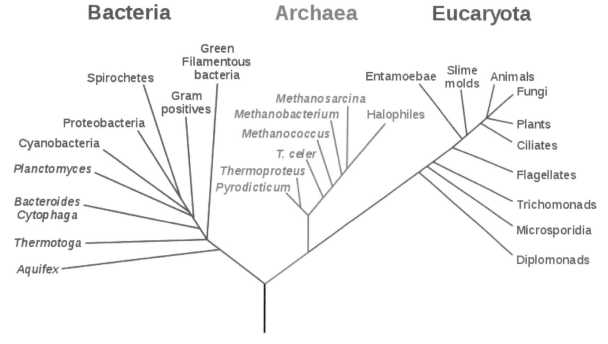

- Phylogenetic tree. Source: NASA and Eric Gaba (from C.R.Woese, O.Kandler and M.L.Wheelis. 1990. *Towards a natural system of organisms: proposal for the domains Archaea, Bacteria, and Eucarya*. Proceedings of the National Academy of Science.)

By tradition:

1) **Parasitology** covers protists and animals;
2) **Mycology** covers Fungi;
3) **Microbiology** covers Monera;
4) **Botany** or **Plant pathology** covers disease-causing plants.

PROTISTS

- Polyphyletic group, includes
    - Sarcomastigophora: Amoebas and Flagellates
    - Apicomplexa: Organisms endowed with a unique organelle, the apical complex
    - Ciliophora: The ciliates
- Asexual and sexual reproduction occurs in these groups
    - In some groups, only asexual reproduction has been observed or is known;
    - In other groups, there is a sexual phase inherent in the life cycle.
- Transmission may be via a fecal-oral route, or via an arthropod vector.

N.B. This is NOT what is meant by the fecal-oral route:
Photograph courtesy of Davida Lassow Margolin, Instructor, Department of New Hampshire, Department of Molecular, Cellular & Biomedical Sciences:

Phylum Sarcomastigophora: The amoebas and flagellates

- Amoebas:
  - Unicellular
  - Most are found in the GI tract
    - Commensals are harmless, but can be easily confused with pathogenic *Entamoeba histolytica*
    - Correct diagnosis is essential
  - Follow fecal-oral route
  - Sexual reproduction is unknown
  - Intestinal forms form metabolically active trophozoites in the intestine, cyst forms for transmission in the environment
- Flagellates
  - Order Kinetoplastida
    - Although sexual reproduction has not been observed under the microscope, there is genetic evidence of genetic exchange
    - High morphological plasticity, depending on whether it is in the invertebrate or vertebrate host
  - Other flagellates show no evidence of genetic exchange
  - Unusual metabolic pathways, especially in *Giardia lamblia*
  - Some intestinal forms form metabolically active trophozoites in the intestine, cyst forms for transmission in the environment
  - *Trichomonas vaginalis* is one of several STD's.

Entamoeba histolytica (amebic dysentery)

Images:
http://www.k-state.edu/parasitology/625tutorials/Ehistolytica.html

Phylogeny:	Superclass Sarcodina
Preferred definitive host:	Humans
Reservoir hosts:	Dogs, Pigs, Monkeys
Vector/intermediate hosts:	None are necessary, but transport by filth flies is possible
Geographical location:	Cosmopolitan
Organs affected:	Coecum, appendix, colon. Advanced disease may include the liver and lungs.
Symptoms and clinical signs:	Mucosal destruction, perforated colons, peritonitis, abscesses in liver, lesions in lungs.
Treatment:	Metronidazole, Dehydroemetine, Chloroquine

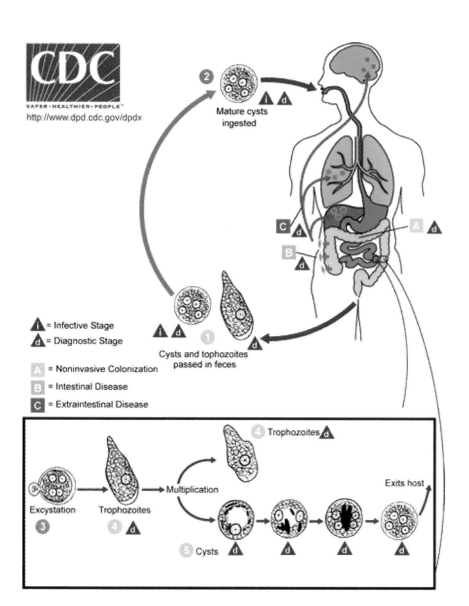

Mature cysts ingested

C
A d
B

= Infective Stage
= Diagnostic Stage

A = Noninvasive Colonization
B = Intestinal Disease
C = Extraintestinal Disease

Cysts and tophozoites passed in feces

Trophozoites

Excystation   Trophozoites   Multiplication   Exits host

Cysts

Entamoeba histolytica trophozoite (1000x)	*Entamoeba histolytica* cyst (1000x)

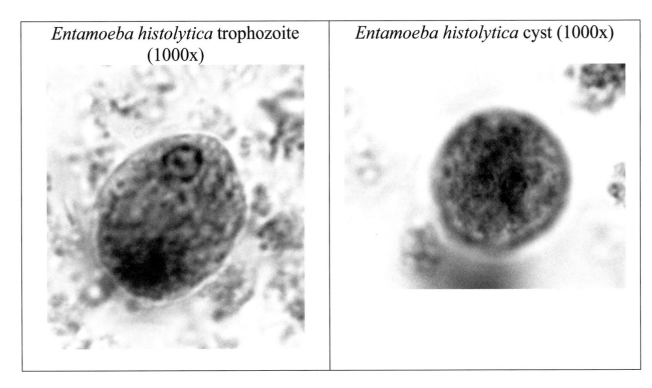

Lesion in the colon caused by *Entamoeba histolytica* (400x)

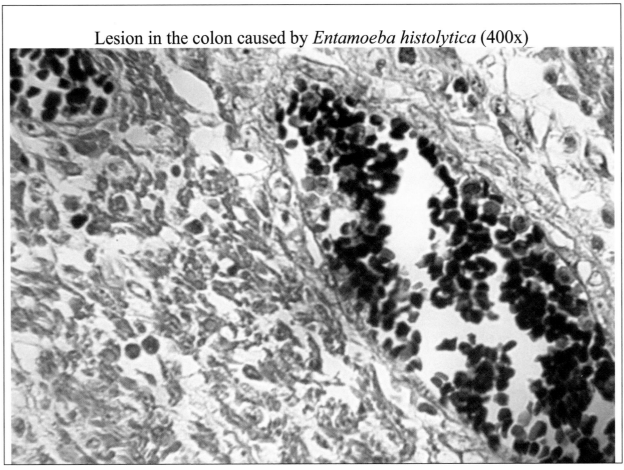

Entamoeba coli

Images:
http://www.k-state.edu/parasitology/625tutorials/Entamoebacoli.html

Phylogeny:	Superclass Sarcodina
Preferred definitive host:	Humans
Reservoir hosts:	None
Vector/intermediate hosts:	None
Geographical location:	Cosmopolitan
Organs affected:	Cecum and general colon
Symptoms and clinical signs:	Symptomless, since E. coli feeds on bacteria, yeast, and on rare occasions, blood cells. This species is frequently mistaken for E. histolytica.
Treatment:	None required

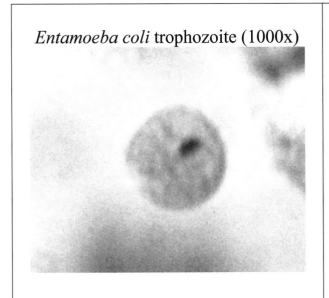

Entamoeba coli trophozoite (1000x)

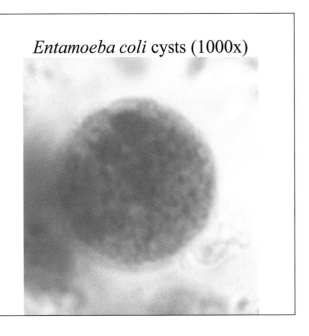

Entamoeba coli cysts (1000x)

Entamoeba gingivalis

Images: http://www.k-state.edu/parasitology/625tutorials/Egingivalis.html

Phylogeny: Superclass Sarcodina

Preferred definitive host: Humans

Reservoir hosts: None, but it will infest primates,
                                   dogs, and cats. Transfer is
                                   possible among avid pet
                                   lovers.

Vector/intermediate host: None

Geographical location: Cosmopolitan

Organs affected: Surface of teeth and gums, gingival
                                   pockets near the base of the
                                   teeth, and sometimes in the
                                   crypts of the tonsils.

Symptoms and clinical signs: None

Treatment: None required

Endolimax nana

Images: http://www.k-state.edu/parasitology/625tutorials/Endolimax.html

Phylogeny	Superclass Sarcodina
Preferred definite host:	Humans
Reservoir hosts:	None
Vector/Intermediate host:	None
Geographical location:	Cosmopolitan
Organs affected:	Lives in the large intestine, mainly at the level of the cecum and feeds on bacteria.
Symptoms and clinical signs:	None. This organism is a commensal which can be confused for pathogenic species
Treatment:	None required

Iodamoeba butschlii

Images: http://www.k-state.edu/parasitology/625tutorials/Iodamoeba.html

Phylogeny:	Superclass Sarcodina
Preferred definitive host:	Humans
Reservoir hosts:	Other primates and pigs
Vector/intermediate host:	None
Geographical location:	Cosmopolitan
Organs affected:	Large intestine, mainly in the cecal area
Symptoms and clinical signs:	Generally none, but in a few cases it has induced ectopic abscesses like those of E. histolytica.
Treatment:	None required

Naegleria fowleri

Images: http://www.k-state.edu/parasitology/625tutorials/Naegleria.html

Phylogeny:	Superclass Sarcodina
Preferred definitive host:	Humans are an accidental host for Naegleria.
Reservoir hosts:	None
Vector/intermediate host:	None
Geographical location:	Cosmopolitan. Cases have been documented in Europe, North America, Africa, New Zealand, and Australia.
Organs affected:	Brain tissue
Symptoms and clinical signs:	Meningoencephalitis, involving convulsions leading to death.
Treatment:	None are available. Infection with Naegleria is always fatal.

Flagellates, other than Order Kinetoplastida

- Most are intestinal,
    - ○ Trophozoites are metabolically active
    - ○ Cysts are formed for dispersal
- Anaerobic metabolism
- Only asexual reproduction has been observed

Giardia lamblia (intestinalis)

Life cycle is similar to that of *Entamoeba histolytica*:

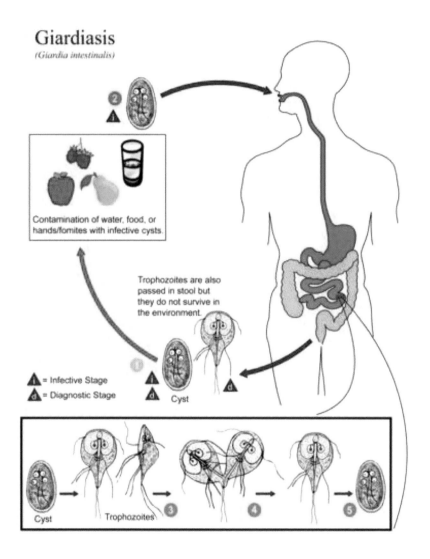

Images:

Cysts: http://www.k-state.edu/parasitology/625tutorials/Protozoa04.html
Trophozoites: http://www.k-state.edu/parasitology/625tutorials/Protozoa02.html

Phylogeny: Order Diplomonadida

Preferred definitive host: Humans

Reservoir hosts: Possibly dogs, cats, rodents, cattle, beaver

Vector/intermediate host: None

Geographical location: Cosmopolitan, but occurs most frequently in warm climates among children.

Organ affected: Duodenum, jejunum, and upper ileum.

Symptoms and clinical signs: Mucus in stools, diarrhea, dehydration, intestinal pain, flatulence, and weight loss.

Treatment: Quinacrine, Metronidazole

Note: European travel agents are advising THEIR customers who arrange visits to the United States, "Don't drink the water! It's infected with G. lamblia!"

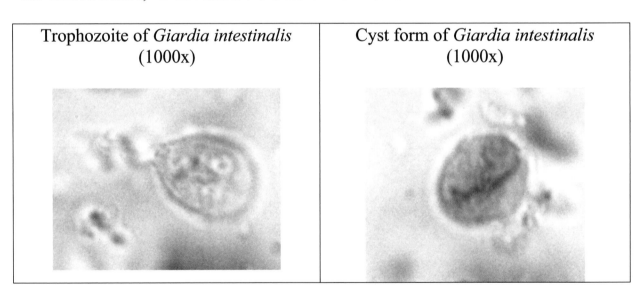

Trophozoite of *Giardia intestinalis* (1000x)	Cyst form of *Giardia intestinalis* (1000x)

Chilomastix mesnili

Images: http://www.k-state.edu/parasitology/625tutorials/Protozoa05.html

Phylogeny:	Order Retortamonadida
Preferred definitive host:	Humans
Reservoir hosts:	Other hosts include chimpanzees, orangutans, monkeys, and pigs
Vector/intermediate host:	None
Geographical location:	Cosmopolitan
Organs affected:	Cecum and colon
Symptoms and clinical sings:	May cause watey stools
Treatment:	None required

Trichomonas vaginalis

Images: http://www.k-state.edu/parasitology/625tutorials/Protozoa01.html

Phylogeny: Order Trichomonadida

Preferred definitive host: Humans

Reservoir hosts: None

Vector/intermediate host: None

Geographical location: Cosmopolitan

Organs Affected: Vagina and urethra of women and in the
                                     prostate, seminal vesicles, and
                                     urethra of men

Symptoms and clinical signs: Frequently symptomless among males,
                                         but some strains cause inflammation,
                                         with itching and a copious white
                                         discharge swarming with
                                         trichomonads. Vaginal secretions
                                         may become greenish and condition
                                         may become chronic and/or recurrent.

Treatment: Metronidazole

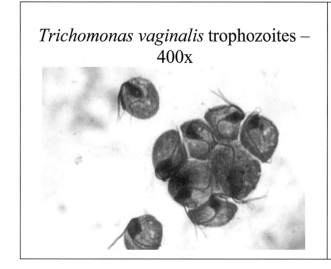

Trichomonas vaginalis trophozoites –
400x

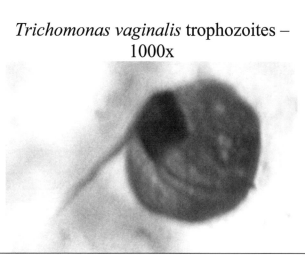

Trichomonas vaginalis trophozoites –
1000x

Order Kinetoplastida

- Part of the Subphylum Mastigophora
- Difficult to treat – drugs containing antimony are the still the standard of care
- High plasticity in morphology among invertebrate and vertebrate hosts. Note differences in morphology in *Leishmania* spp. below:

Generalized life cycle for *Leishmania* spp.

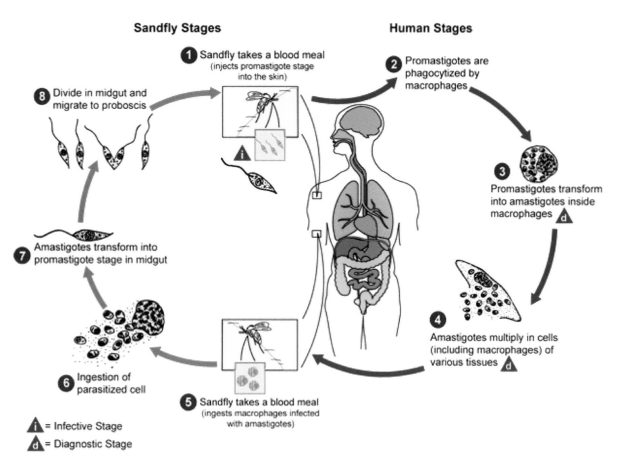

Leishmania donovani

Images: http://www.k-state.edu/parasitology/625tutorials/Kinetoplastids01.html

Culture (promastigote) and endoparasitic (amastigote) forms of all species appear similar under the light microscope:

Culture forms of *Leishmania* spp. (1000x)	Endoparasitic forms of *Leishmania* spp. (1000x)
	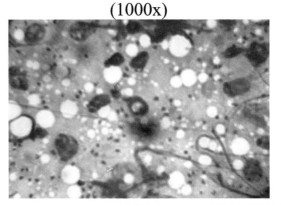

Phylogeny: Order Kinetoplastida

Preferred definitive host: Humans

Reservoir hosts: Dogs, jackals, foxes

Intermediate/vector hosts: *Phlebotomus* spp. sandflies

Geographical location: Southern Russia, China, Northeast India,
                                        Bangladesh, Central and South America

Organs affected: Reticuloendothelial system

Symptoms: Fever, anemia, edema, difficulty breathing,
                                        diarrhea, emaciation, hepatosplenomegaly
                                        as compensation for anemia

Treatment: Antimony sodium gluconate, Pentamidine

Leishmania tropica

Rhinophymous Leishmaniasis: A New Variant – from Medscape Infectious
Diseases
http://www.medscape.com/viewarticle/704662?src=mp&spon=3&uac=40240FX

Phylogeny:	Order Kinetoplastida
Preferred definitive host:	Humans
Reservoir hosts:	Dogs, rodents
Intermediate/vector hosts:	*Phlebotomus* spp. sandflies
Geographical location:	West-Central Africa, Mediterranean region, India, South America, Central America, Ethiopia
Organs affected:	Reticuloendothelial system, skin
Symptoms:	Ulcers and sores on skin
Treatment:	Antimony sodium gluconate. Frequently self-healing with lasting immunity.

Leishmania braziliensis

Images:

Phylogeny:	Order Kinetoplastida
Preferred definitive host:	Humans
Reservoir hosts:	Dogs, rodents, cats, kinkajou
Intermediate/vector hosts:	*Lutzomyia* spp. sandflies
Geographical location:	Central and South America
Organs affected:	Nasal system and buccal mucosa
Symptoms:	Destruction of cartilaginous and soft tissue. Ulceration of lips, palate, pharynx leading to deformity.
Treatment:	Antimony sodium gluconate, Amphotericin B, cycloguanil pamoate

Trypanosoma gambiense and *T. rhodesiense* (African trypanosomiasis)

Life cycle of African (sleeping sickness) species:

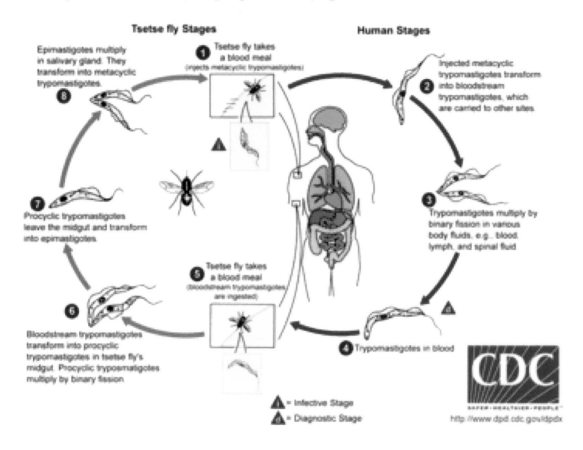

Images: http://www.k-state.edu/parasitology/625tutorials/Protozoa06.html

Phylogeny:	Order Kinetoplastida
Preferred definitive host:	Humans
Vector/intermediate host:	Tsetse flies (genus *Glossina* spp.)
Geographical location:	Central and East central Africa
Organs affected:	Blood, central nervous system.
Symptoms and clinical signs:	Lymph nodes swell, increasing apathy, mental dullness, tremor of the tongue, hands and trunk, anemia

due to lysis of rbc's, somnambulism.

Treatment: Arsenic drugs, suramin, pentamidine, Berenil.

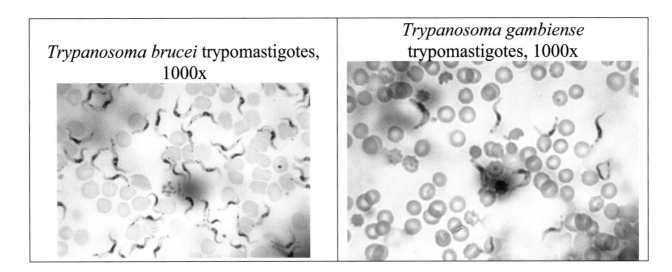

| *Trypanosoma brucei* trypomastigotes, 1000x | *Trypanosoma gambiense* trypomastigotes, 1000x |

Trypanosoma cruzi (American trypanosomiasis - Chagas' Disease)

Images: http://www.k-state.edu/parasitology/625tutorials/Kinetoplastids01.html

Life cycle of American trypanosomiasis:

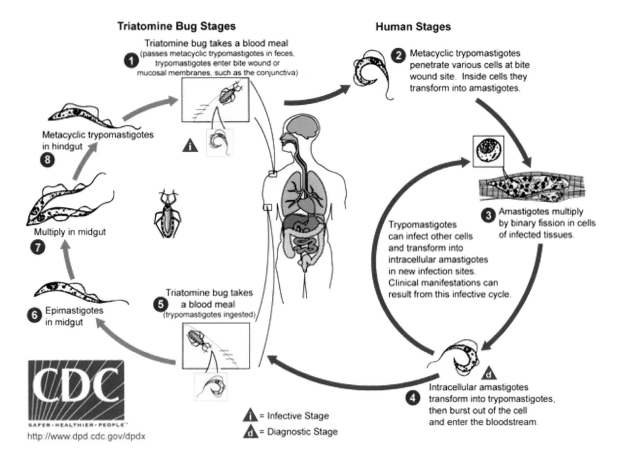

Phylogeny: Order Kinetoplastida

Preferred definitive hosts: Humans

Reservoir hosts: Dogs, cast, opossums, armadillos, wood rat

Intermediate/vector hosts: *Triatoma* bugs in Uruguay, Chili, Argentina, *Rhodnius prolixus* in northern South America and Central America

Geographical location: Central and South America

Organs affected: Lymph node, nervous tissue, heart muscle

Symptoms and clinical signs: Swelling of lymph nodes, progressive deterioration of nervous tissue, resulting in loss of strength, nervous disorders, heart failure, megaesophagus or megacolon

Treatment: No effective drug

Trypanosoma cruzi trypomastigote form, 1000x (Note distinctive 'C' shape)	*Trypanosoma cruzi* amastigotes in cardiac tissue.
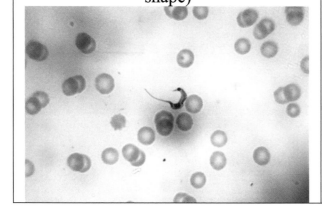	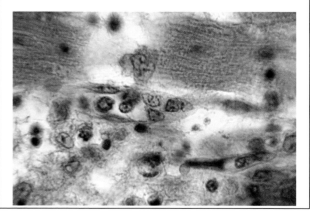

Phylum Apicomplexa

- Asexual and sexual phases alternate
- Sexual phase often in a vertebrate, except for *Plasmodium* spp., causative agent for malaria
- Apical complex is located in anterior part of the cell: An organ complex of the Apicomplexa that appears as a conical structures on the tapered end (or the apical end) of the cell, and contains rhoptries, micronemes, polar rings, and conoid. The apical complex is said to help the apicomplexan when invading an animal cell.

Plasmodium spp., including

P. falciparum, P. malariae, P. ovale, and *P. vivax* (malaria)

Images:
Plasmodium falciparum:
http://www.k-state.edu/parasitology/625tutorials/Plasmodium01.html
Plasmodium lophurae exflagellation:
http://www.k-state.edu/parasitology/625tutorials/Apicomplexa05.html
Plasmodium malariae:
http://www.k-state.edu/parasitology/625tutorials/Plasmodium04.html
http://www.cbu.edu/~seisen/Malaria/index.htm

Generalized life cycle of *Plasmodium* spp.

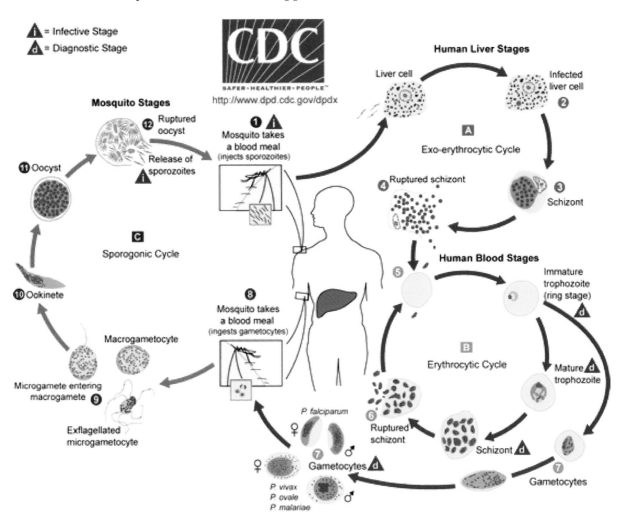

Plasmodium falciparum blood stages, 1000x	*Plasmodium vivax* blood stages, 1000x
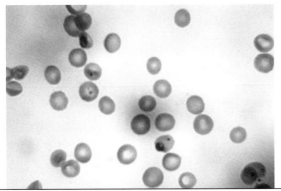	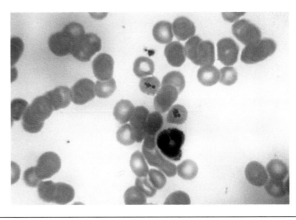
Plasmodium spp. oocysts, c.s., 100x	*Plasmodium* spp. oocysts, 1000x, showing individual sporozoites
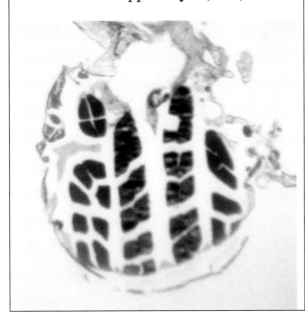	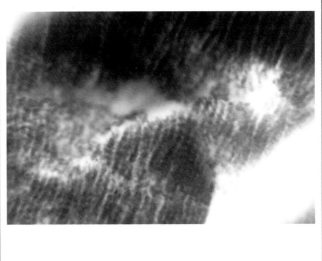

Timeline, adapted from Maher, B.A. (2005). Fever Pitch. The Scientist 18(10):25.

Year	Discovery
400 BC	Susruta, A Brahmin priest, describes malarial fever that he attributes to mosquito bites.
95 BC	Lucretius suggests that an organism rather than poisonous air or miasma might cause malaria, which means "bad air" in Italian.
450 AD	A widespread epidemic occurs in Lugnano, north of Rome, according to forensic DNA evidence.
1638	Juan del Vego uses a tincture from bark of a tree to treat the Countess of Chinchon in Peru; the remedy is later named quinine.

1716	Giovanni Maria Lancisi, physician to three popes, notes that draining swamps curbs malaria; he suggests an insect origin.
1880	French army surgeon Charles Louis Alfonse Laveran identifies malaria parasite; wins Nobel Prize in 1907.
1894	Patrick Manson hypothesizes that an external vector transmits malaria.
1897	Ronald Ross, military physician in India, observes malaria parasite in Anopheline mosquito guts, wins Nobel Prize in 1902.
1934	Chemist Hans Andersag at Bayer laboratories in Germany discovers chloroquine (resochin), but the compound is largely forgotten. It won't be recognized as a safe effective antimalarial drug until 1946.
1939	Paul Muller in Switzerland notes insecticidal properties of DDT, synthesized nearly a 100 years earlier by Othmar Zeidler, a German chemistry student.
1947-1951	National Malaria Eradication Program established by state and federal agencies essentially eradicates malaria in the United States.
1956	World Health Organization (WHO) launches Global Malaria Eradication Program.
1960's	Widespread drug-resistant parasites and DDT-resistant mosquitoes are noted
1962	Rachel Carson publishes *Silent Spring,* about the environmental effects of DDT.
1967	WHO abandons malaria eradication in favor of control.
1972	The US Environmental Protection Agency bans the use of DDT
1979	Chinese researchers describe artemisinin, a wormword-derived treatment noted in ancient texts.
1983	First *Plasmodium* gene is cloned
1998	WHO initiates Roll Back Malaria program with the goal of halving the burden of malaria by 2010.
2000	UK researchers produce the first transgenic mosquitoes.
2002	International consortia publish the sequence of *Plasmodium falciparum* and a draft sequence of *Anopheles gambiae*.

There is a long association of humans with malaria. There are 3,000 year-old records from India and China which describe the classic symptoms of chills and fever associated with the erythrocytic stage of malaria.

An Excerpt from Shakespeare's Julius Caesar and how it relates to the medieval perception of the evils and diseases which await someone who dares to walk out into the dank, dark morning...

Act II, Scene I: Brutus has decided to join the conspiracy to assassinate Julius Caesar. His conscience is obviously bothering him, and he is pacing the floor of his courtyard in the early morning hours before sunrise when his wife, Portia, startles him...

Enter Portia.

Por. Brutus, my lord!

Bru. Portia, what mean you? Wherefore rise
you now?
It is not for your health to commit
Your weak condition to the **raw cold morning.**

Por. Now for yours neither. You have ungently, Brutus,
Stole from my bed: and yesternight, at supper,
You suddenly arose, and walk'd about,
Musing and sighing, with your arms across;
And when I ask'd you what the matter was,
You star'd upon me with ungentle looks:
I urg'd you further; then you scratch'd your head,
And too impatiently stamp'd with your foot:
Yet I insisted, yet you answer'd not;
But with an angry wafture of your hand
Gave sign for me to leave you: so I did;
Fearing to strengthen that impatience
Which seem'd too much enkindled; and withal
Hoping it was but an effect of humour,
Which sometime hath his hour with every man.
It will not let you eat, nor talk, nor sleep;
And, could it work so much upon your shape
As it hath much prevail'd upon your condition,
I should not know you, Brutus. Dear my lord,
Make me acquainted with your cause of grief.

Bru. I am not well in health, and that is all.

Por. Brutus is wise, and were he not in health,
He would embrace the means to come by it.

Bru. Why, so I do.- Good Portia, go to bed.

Por. Is Brutus sick? And is it physical
To walk unbraced, **and suck up the humours
Of the dank morning?** What, is Brutus sick, -
And will he steal out of his wholesome bed,
**To dare the vile contagion of the night,
And tempt the rheumy and unpurg'd air
To add unto his sickness?**...

(*Editor note: Actually, Portia is much more perceptive than that. Her NEXT line is, "No, my Brutus: You have some sick offence within your mind, Which by the right and virtue of my place I ought to know of." A few lines later, she says, "...for here have been Some six or seven [men], who did hide their faces Even from darkness.*)

Phylogeny:	Subphylum Apicomplexa
Preferred definitive host:	Technically, mosquitos are the definitive host since the parasite undergoes sexual reproduction in the mosquito. By convention, mosquitos are considered the "vectors" to humans.
Reservoir hosts:	None
Vector/intermediate host:	Mosquitos, particularly those of the genus *Anopheles*.
Geographical location:	Central and South America, Africa, Middle East, Asia, Pacific Islands
Organs affected:	Liver, blood, kidney
Symptoms and clinical signs:	Most symptoms are associated with

its effects on erythrocytes.
Symptoms commonly include chills,
fever, and anemia. Other symptoms
include muscle pain, headache, loss
of appetite, nausea, vomiting,
jaundice, and renal failure.

Treatment: Chloroquine, Primaquine, Sulfamethoxine,
                                          Pyrimethamine, Sulfadiazine, Quinine,
                                          Amodiaquine.

Groovy Web site(s):   Malaria Foundation International
                                  http://www.malaria.org

Note: A number of genetic conditions have evolved among human populations in response to malaria. The best known are sickle-cell anemia and favism, a deficiency of glucose-6-phosphate dehydrogenase.

Some drugs used in the treatment of malaria are nasty, and have psychological effects. Here is the text of an e-mail distributed, requesting information regarding Lariam:

From: "Dan Olmsted" <DOlmsted@upi.com
To: <info@rpcv-wa.org>
 Sent: Wednesday, June 05, 2002 1:14 PM
Subject: Lariam query from UPI

We would appreciate it if you could post this and/or send to volunteers. If you have any questions feel free to contact me at 202 302 3753 or via e-mail. Thanks, Dan Olmsted, Washington Bureau Chief, United Press International. United Press International is investigating the anti-malaria drug Lariam and is interested in hearing from Peace Corps volunteers about any problems they may have experienced.

If you experienced psychiatric or other reactions to the drug either during or after your Peace Corps years, we would like to hear from you. We also are interested in hearing about any reports of volunteers not taking the drug because of side effects; what kind of warnings you received; whether your complaints about side effects were taken seriously, and how Peace Corps medical officers dealt with the issue of side effects. We also would like to find former medical officers or Peace Corps officials who would talk to us. Also, we are interested in any information about suicidal thinking or behavior, or actual suicides or unexplained deaths, that might be connected with the drug. UPI published an article on side effects including suicide on May 21; you can read it by going to UPI.com and typing in Lariam, or going to Newsday.com and doing the same thing (that is a shorter version).

You can e-mail me at dolmsted@upi.com. Please include a phone number and indicate whether you would be willing to be quoted by name (we only use named sources in our reporting). Also, if you are attending the Peace Corps convention in Washington in June, please let us know.

We are taking the issue of side effects very seriously and are committed to full and accurate reporting about the situation.

Sincerely,

Dan Olmsted
Washington Bureau Chief
United Press International

'Happy' **Malaria Awareness Day**!

Today on campuses around the country students with the Student Campaign for Child Survival are demanding concrete US action to prevent the deadly impact of malaria. On lawns, in their homes, in dining halls, around the dinner table and as a part of the **"Malaria Bites, Bite Back"** nation-wide house party students are coming together, preparing foods from malaria regions, learning about the devastating impact of the disease, and sending a bold message to their elected officials in DCs – **WE ARE BITING BACK!**

- **Students from California to New York, from Wisconsin to Texas will be taking part.**
- **Stanford**: Students will be tabling on campus, hosting a speaker and screening a film – all on **African malaria**.
- **Cornell**: Campaign members are hosting house parties and taking pictures of all the partygoers who are **'biting back.'**

- **St. Scholastica**: Students participating in the Mayfest **Fun Run** will be sporting **"I'm Biting Back"** stickers. Non-runners can pose for pictures in the student union or make a fee call to their elected officials in DC.
- **Beloit College**: At this small school in Southern Wisconsin students are taking over their entire (and only) dining hall with a school-wide game of **Malaria Trivia**.
- **Texas A&M**: Students are gathering for a more intimate **roundtable discussion** on the US response to the deadly disease.
- **And many more!**

Photos of students participating in their events and holding signs that read "I'm biting back" will be compiled by students at SCCS's DC headquarters and presented to Representative Obey (D-WI) with our semester's primary request - fully fund the Global Fund and PMI.

Our press release is attached, updates on how everything panned out will follow, and the house-party action kit, if you are curious, is still online here.

Keep up the good work everyone,

Simon Stumpf | SCCS National Organizer
The Student Campaign for Child Survival
c: 320 420 0959 | supportchildsurvival.org

CHARACTERISTICS OF *PLASMODIUM* SPP.

PARAMETER	VIVAX	FALCIPARUM	OVALE	MALARIAE
CIRCADIAN CYCLE OF FEVERS	48 hours	IRREGULAR - 48 hours	72 hours	72 hours
OCCURRENCE	Temperate zone & North Africa & Vietnam	Tropical: Accounts For 50% of cases	Africa, S.E. Asia, New World	Tropics: Java & New Guinea
CELLULAR MARKINGS	Schuffner's Dots	Maurer's Cleft	Schuffner's Dots	Absent
EXOERYTH-ROCYTIC GENERA-TIONS	Several	Only 1	?	Relapses Possible
AGE OF SUSCEP-TIBLE RBC'S	Only young	Any age	Aging	Any age, but low incidence
# MEROZOITES	16	16	8	8
MULTIPLE INFECTIONS OF RBC'S?	Rare	Frequent	No	No
PROTECTION BY SICKLE CELL TRAIT	No	Yes	No	No
NECESSITY OF DUFFY FACTORS	Yes	No	No	No

Toxoplasma gondii

Brain cyst:
http://www.k-state.edu/parasitology/625tutorials/Cysts01.html
Live brain cyst:
http://www.k-state.edu/parasitology/625tutorials/Apicomplexa05.html

Life cycle:

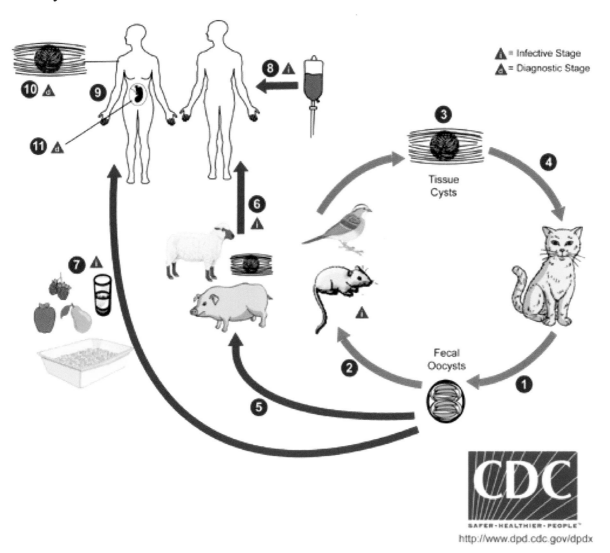

Toxoplasma gondii, from culture, 1000x	*Toxoplasma gondii* in brain c.s., 1000x

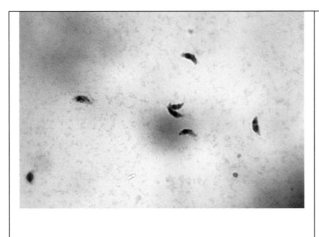 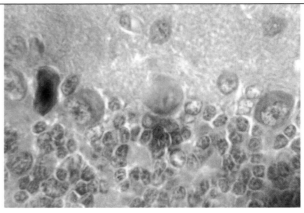

Phylogeny: Subphylum Apicomplexa

Preferred definitive host: Domestic cats, Puma, Ocelot,
                                bobcat, Jaguarundi

Reservoir hosts: Technically none, but cockroaches, flies
                                and leeches serve as transport
                                hosts.

Vector/intermediate host: Humans, Domestic animals such as
                                sheep, wild animals such as sheep,
                                insectivores, rodents, pigs,
                                herbivores.

Geographical location: Cosmopolitan

Organs affected: Lymph glands, lung, liver, heart, brain,
                                eyes. *Toxoplasma* can pass through
                                the placental barrier and affect
                                the developing fetus.

Symptoms and
clinical signs: Among adult humans, it can cause fever, headache,
                                  muscle pain, anemia, spastic paralysis,
                                  blindness, myocarditis, permanent heart
                                  damage. Infection among pregnant women

may cause stillbirths or spontaneous abortions. Congenital conditions include hydrocephalus, microcephaly, cerebral calcification, chorioretinitis and psychomotor disturbances.

Treatment: Pyrimethamine with trisulfapyrimidines.

Cryptosporidium spp.

Images:
Life cycle:
http://www.k-state.edu/parasitology/625tutorials/Crypto01.html
Stages:
http://www.k-state.edu/parasitology/625tutorials/Apicomplexa07.html

Phylogeny:	Phylum Apicomplexa
Preferred definitive host:	Difficult to determine since there are 10 named species among humans, birds, and other mammals.
Reservoir hosts:	Oocysts taken from an immunodeficient person were used to infect kittens, puppies and goats.
Vector/intermediate host:	None
Geographical location:	Cosmopolitan
Organs affected:	Small intestine
Symptoms and Clinical signs:	Among immunocompetent individuals, it causes a self-limiting diarrhea and abdominal cramps lasting 1 to 10 days. However, it causes a profuse, watery diarrhea among immunosuppressed (AIDS) which can persist for months and be life-threatening.
Treatment:	No effective drug treatment has been found yet.

Phylum Ciliophora – the ciliates

- Sexual reproduction is an inherent part of the ciliate life cycle, observed in free-living specimens, as in *Paramecium* spp.
- Cilia used in locomotion

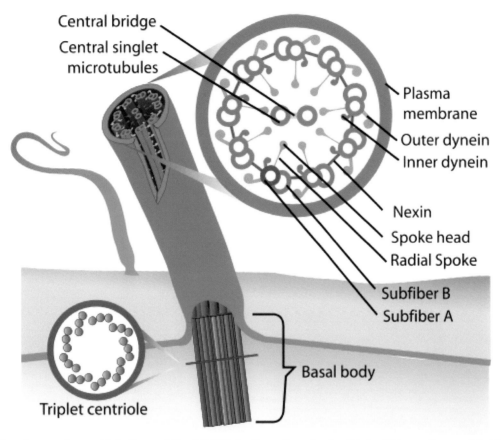

Balantidium coli

Images:

Trophozoites:
http://www.k-state.edu/parasitology/625tutorials/Balantidium.html
Cysts:
http://www.k-state.edu/parasitology/625tutorials/Ciliates.html

Phylogeny:	Phylum Ciliophora
Preferred definitive host:	Humans
Reservoir hosts:	Pigs, guinea pigs, rats, other mammals.
Intermediate/vector hosts:	None
Geographical location:	Most common in Philippines, but is cosmopolitan
Organs affected:	Cecum and colon
Symptoms: epithelium	Proteolytic enzymes digest the intestinal
	of the host. Ulcer is flask-shaped, and causes lymphocytic infiltration, hemorrhage, secondary bacterial infection. Large intestine and appendix may be perforated.
Treatment:	Carbarsone, diiodohydroxyquin, tetracycline. Epidemiological control and treatment are similar to those of *E. histolytica*.

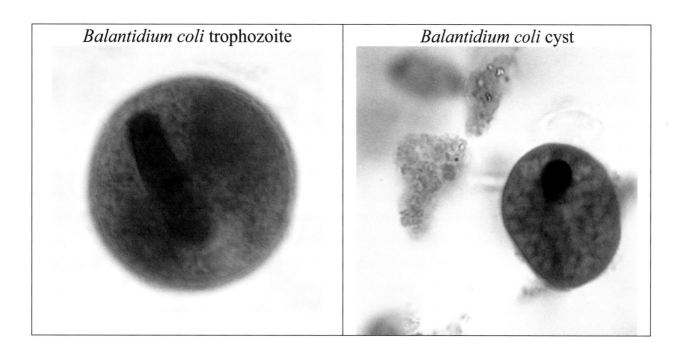

| *Balantidium coli* trophozoite | *Balantidium coli* cyst |

PLAYTHELMINTHES

Fasciola hepatica cross-section, 100x

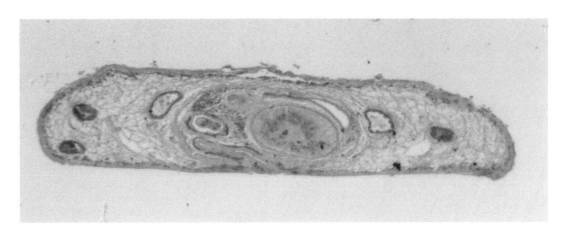

Planaria c.s., 40x

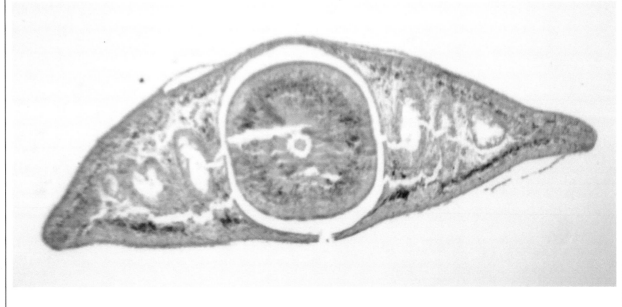

- Triploblastic
- Acoelomate
- Dorso-ventrally flattened
- Digestive system, when present, is a blind sac

- Respiratory and circulatory systems lacking; gastrovascular cavity takes that function
- Excretory system includes protonephridia, with flame cells generating water currents for osmoregulation
- Parasitic forms generally lack pigments and eye spots
- Most are hermaphroditic
- Includes 5 major classes
  - Turbellaria
  - Aspidogastrea
  - Cestoidea
  - Monogenea
  - Digenea
- Turbellaria
  - Free-living and commensals
  - Freshwater and marine habitats
  - *Dugesia* (planaria) is an example
- Aspidogastrea
  - Very primitive
  - Adults found in pericardial cavity of freshwater mollusks
  - Point to a very ancient association with mollusks
- Monogenea
  - External parasites of aquatic vertebrates
    - Skin, and gills of fish
  - Require one host for completion of life cycle
  - Relatively small size

Class Digenea

- Generally hermaphroditic
- Require 2 intermediate hosts

THE PHENOMENON OF POLYEMBRYONY IN DIGENETIC TREMATODES

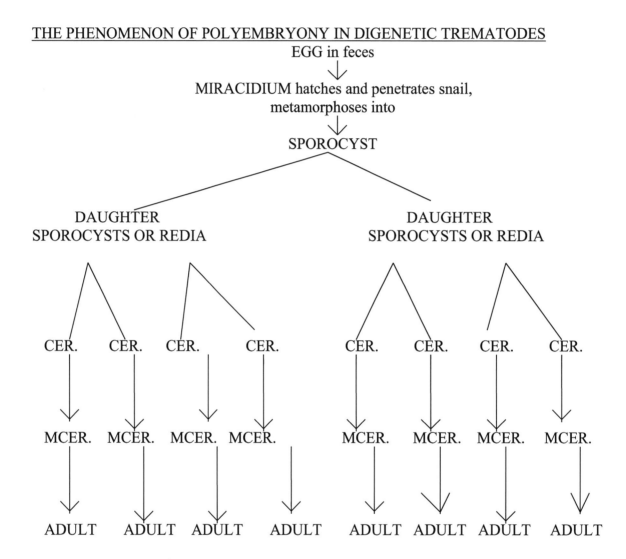

N.B. CERC = Cercariae

MCERC = Metacercariae

Any or all of the following life stages may be present:

Daughter sporocysts or redia

If species has redia following mother sporocyst, then a generation of Daughter Redia is possible

Metacercariae

Clonorchis sinensis (Chinese liver fluke)

Images:

Adult:
http://www.k-state.edu/parasitology/625tutorials/Clonor01.html
Eggs:
http://www.k-state.edu/parasitology/625tutorials/Clonor02.html
Comparison of preserved specimen and line drawing:
http://www.k-state.edu/parasitology/625tutorials/Trematodes08.html

Life cycle:

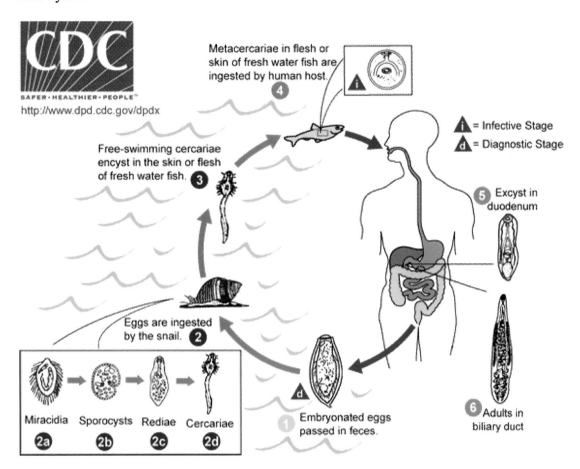

PHYLOGENY: Subclass Digenea, Order Opisthorchiata

Preferred definitive host: Humans

Reservoir hosts: Dogs and cats are probably most important.

Others may include pigs, rats, and camels.

Vector/intermediate host: #1. Snail-Genus *Parafossarulus manchouricus*:
                                 #2. Fish-mostly cyprinids.

Geographical location: Japan, Korea, Taiwan, Vietnam

Organs affected: Bile duct and liver

Symptoms and clinical signs: Erosion of epithelial lining and fibrosis of the liver
                                 occur. Symptoms include ascites, bile
                                 retention, gallstone formation, indigestion,
                                 diarrhea, and hepatomegaly.

Treatment: Praziquantel.

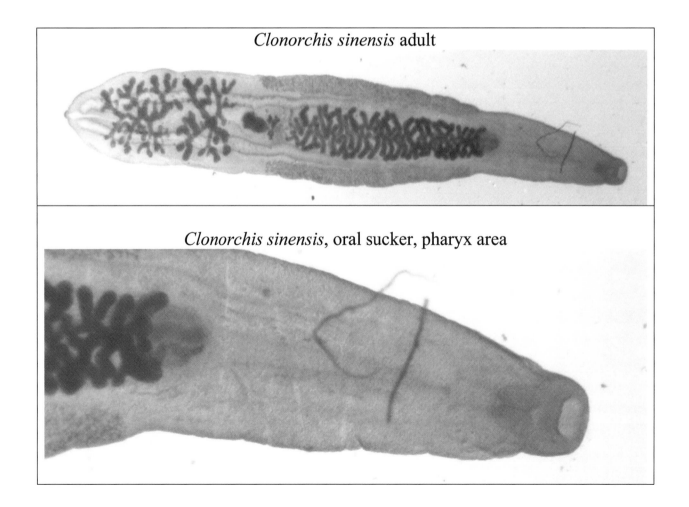

Clonorchis sinensis adult

Clonorchis sinensis, oral sucker, pharyx area

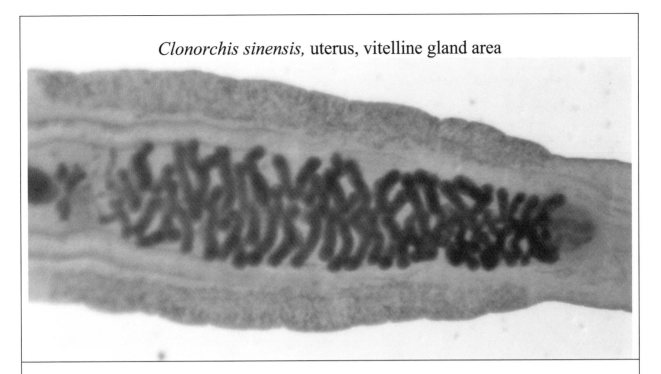

Clonorchis sinensis, uterus, vitelline gland area

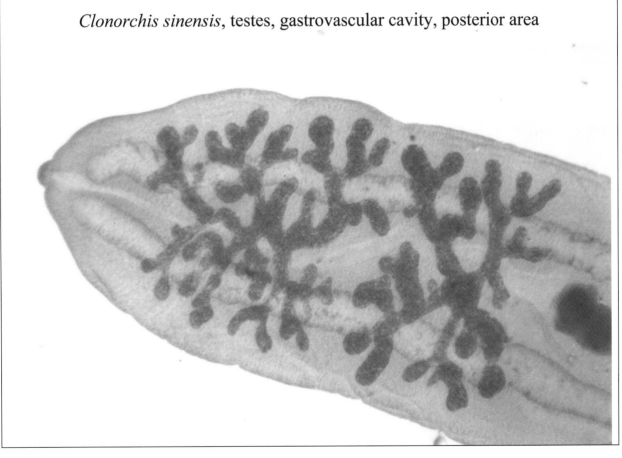

Clonorchis sinensis, testes, gastrovascular cavity, posterior area

Clonorchis sinensis adults, c.s., *in situ*

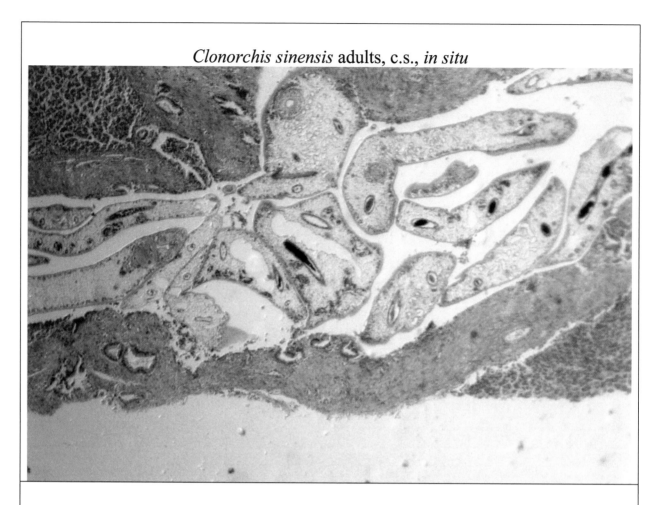

Clonorchis sinensis ova, 1000x, (Note operculum at one end.)

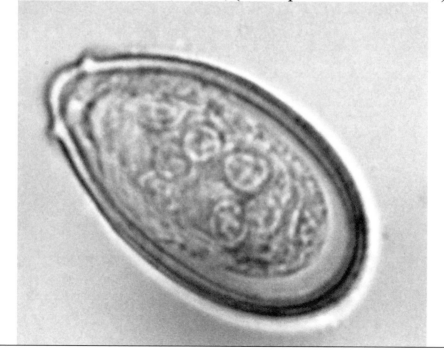

Fasciola hepatica (Sheep liver fluke)

Images:

Adult:
http://www.k-state.edu/parasitology/625tutorials/Fasciola01.html
More adults:
http://www.k-state.edu/parasitology/625tutorials/Hepatica.html
Eggs:
http://www.k-state.edu/parasitology/625tutorials/Platys01.html
More eggs:
http://www.k-state.edu/parasitology/625tutorials/Fasciola02.html

Life cycle:

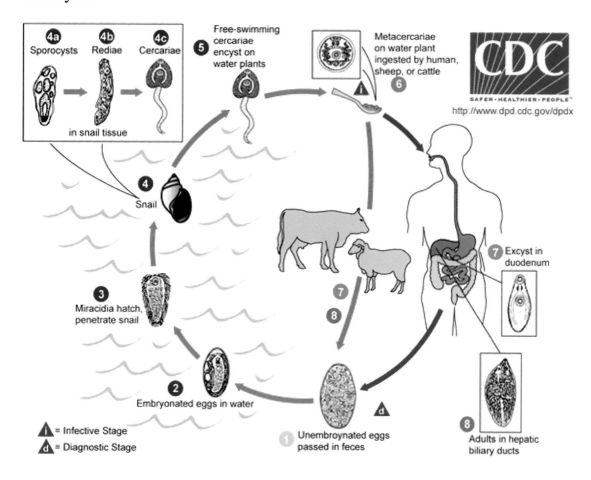

Phylogeny: Subclass Digenea, Order Echinostomata

Preferred definitive host:	Sheep and cattle, rarely among humans
Reservoir hosts:	Sheep, cattle, rabbits
Vector/intermediate hosts:	#1. Snails – *Fossaria modicella* or *Stagnicola bulimoides*; #2. Metacercariae encyst on vegetation.
Geographical location:	Cosmopolitan. Human cases documented in Central & South America, Africa, Asia Europe
Organs affected:	Biliary ducts, liver.
Symptoms:	Necrosis of liver occurs because of migration through the liver. Anemia can result in heavy infections. Worms in bile ducts cause inflammation and edema, leading to fibrous tissue forming in walls of the ducts. Back pressure causes atrophy of liver parenchyma, thus leading to cirrhosis and jaundice. Ectopic infections occur in eye, brain, skin and lungs.
Treatment:	Rafoxanide, praziquantel

Fasciola hepatica adult

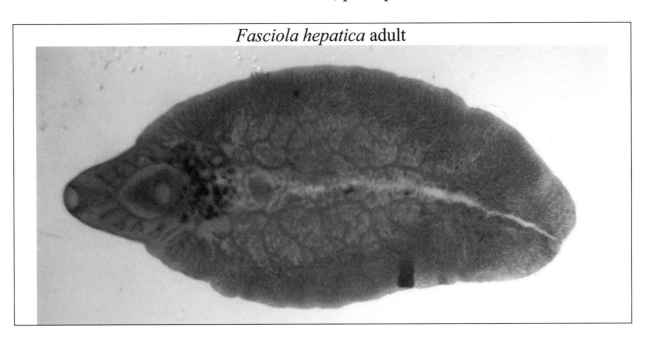

Fasciola hepatica, oral sucker, pharynx, ventral sucker, ova

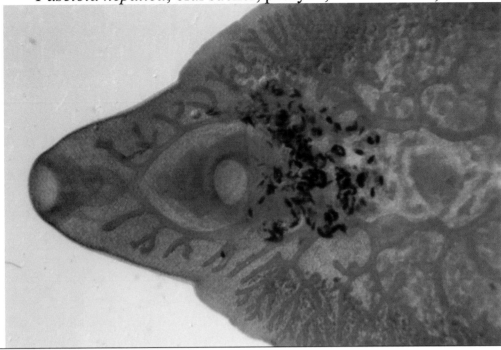

Fasciola hepatica adult, posterior

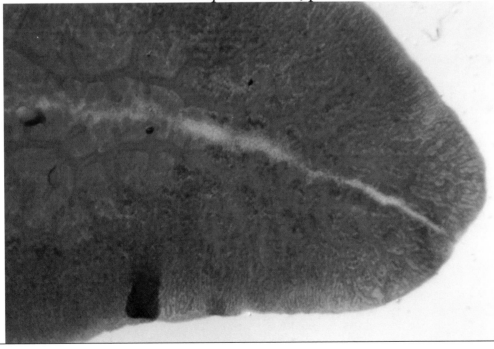

Fasciola hepatica redia

Fasciola hepatica cercariae

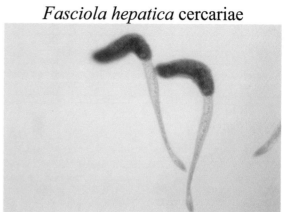

Fasciola hepatica ovum, 400x

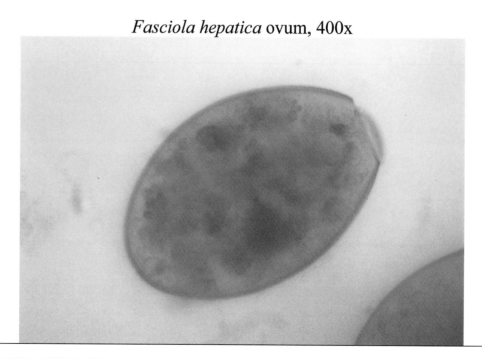

Fasciolopsis buski

Images:
```
 Adults:
```
http://www.k-state.edu/parasitology/625tutorials/Buski.html
```
Comparison of preserved specimen with line drawing:
```
http://www.k-state.edu/parasitology/625tutorials/Trematodes09.html

Life cycle:

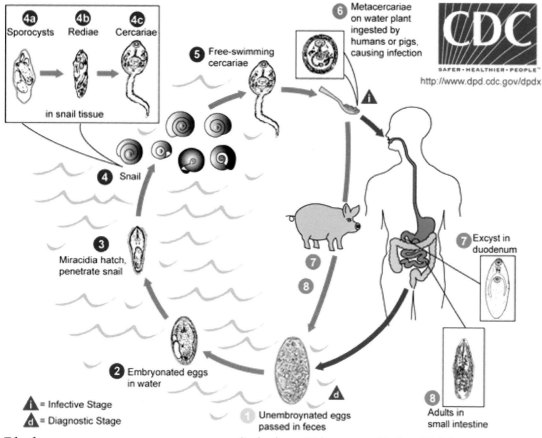

Phylogeny:	Subclass Digenea, Order Echinostomata
Preferred definitive host:	Humans
Reservoir hosts:	Pigs
Vector/intermediate host:	Snails, genera *Segmentina* or *Hippeutis*
Geographical location:	China and Southeast Asia

Organs affected: Small intestine

Symptoms and
    Clinical signs: Blockage of passageway will cause ulceration, hemorrhage, abscesses, hepatic fibrosis, and verminous intoxication.

Treatment: Praziquantel

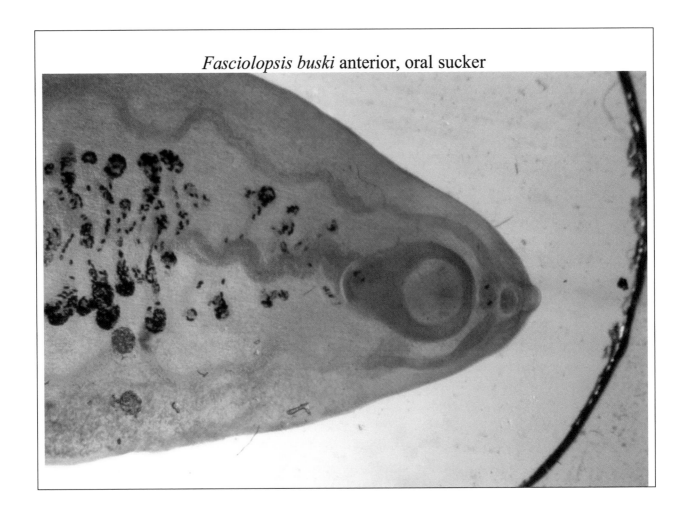

Fasciolopsis buski anterior, oral sucker

Fasciolopsis buski middle

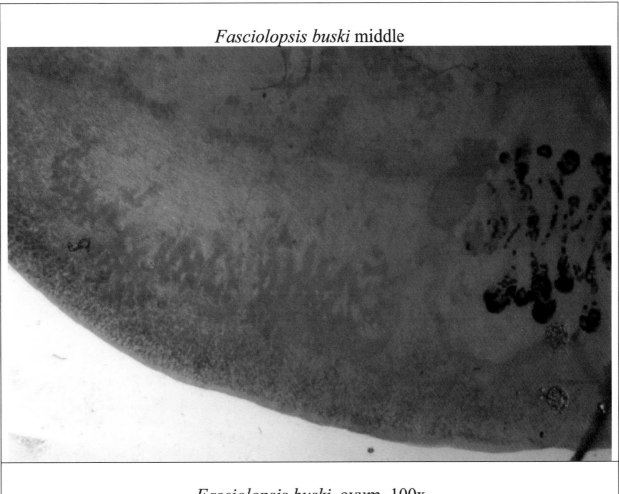

Fasciolopsis buski, ovum, 100x

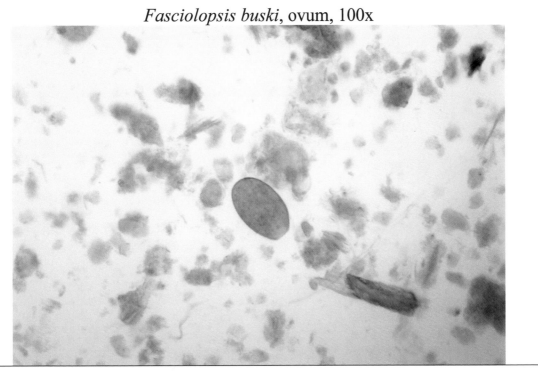

Fasciolopsis buski ovum,

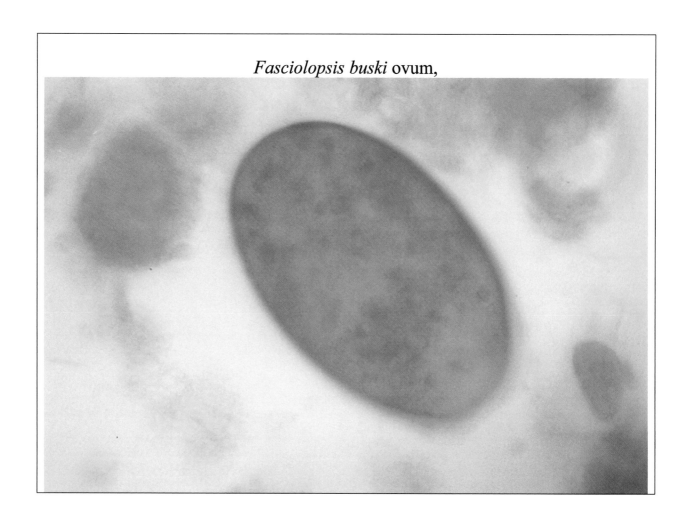

Paragonimus westermani (Chinese lung fluke)

Images:

Adult *Paragonimus kellicotti*
http://www.k-state.edu/parasitology/625tutorials/Paragon01.html
Eggs:
http://www.k-state.edu/parasitology/625tutorials/Paragon02.html

Life cycle:

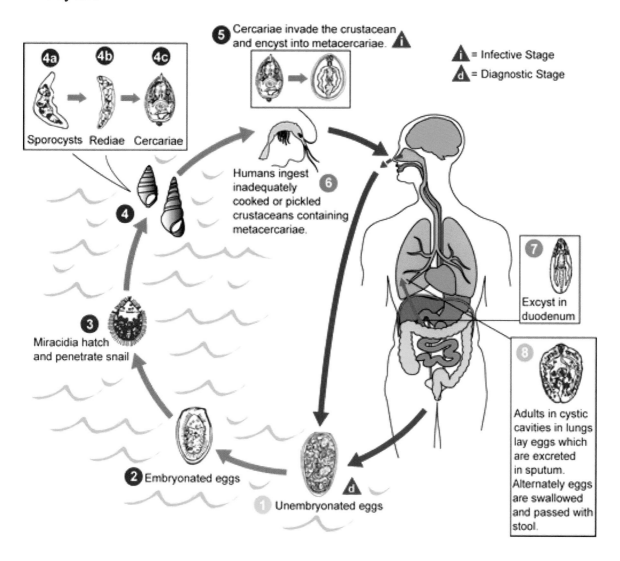

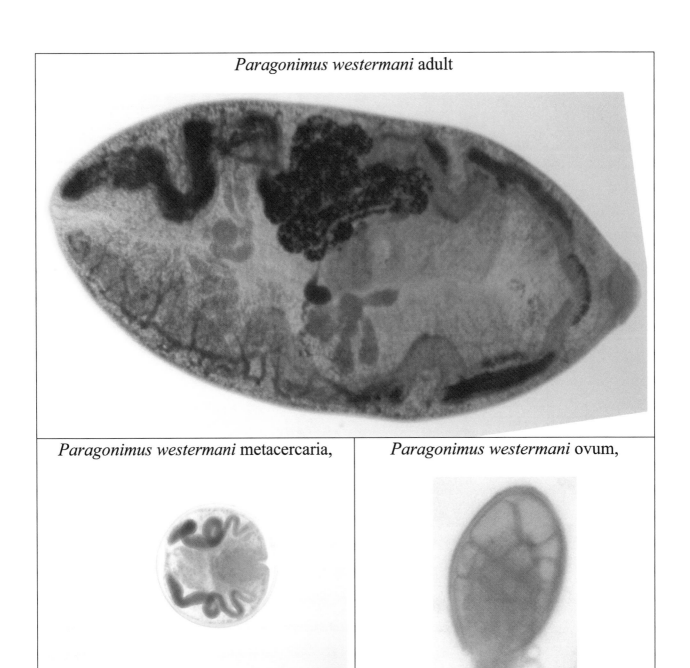

Paragonimus westermani adult

Paragonimus westermani metacercaria,

Paragonimus westermani ovum,

Phylogeny: Subclass Digenea, Order Plagiorchiata

Preferred definitive host: Carnivores (e.g. felids, canids, viverids, and mustelids), rodents, and pigs.

Reservoir hosts: Humans

Vector/intermediate hosts: 1. Snail of Family Thieridae; 2. Crab-*Eriocheir japonicus.*

Geographical location: Japan, Korea, Taiwan, Western Africa, South America

Organs affected: Mainly the bronchioles of the lungs, but the worms may wander into the brain or mesentery.

Symptoms and
clinical signs: Victim suffers from breathing difficulties and chronic cough. Worm is often fatal due to penetration of the brain, spinal cord, or heart.

Treatment: Bithionol, Praziquantel

Schistosoma spp. (Blood flukes)

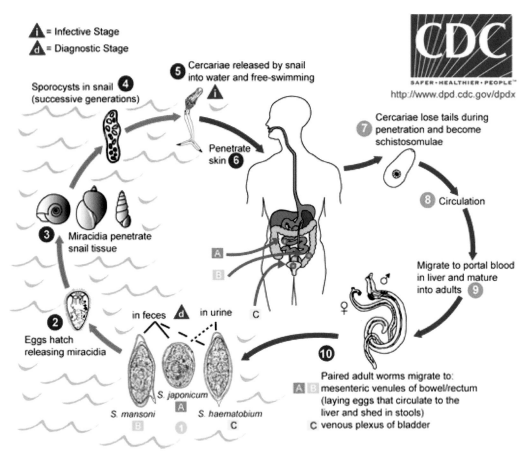

Comparison of *Schistosoma* spp. ova		
S. mansoni (lateral spine), 1000x	*S. haematobium* (terminal spine), 1000x	*S. japonicum* (rudimentary spine), 1000x

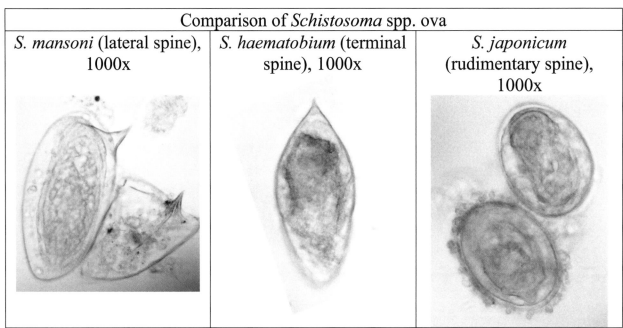

Schistosoma mansoni miracidium, 1000x	*Schistosoma mansoni* cercariae, 1000x. Note forked tail, characteristic of schistosome cercariae.
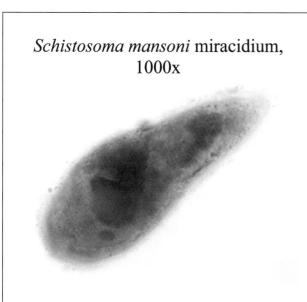	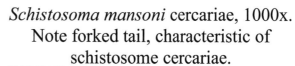 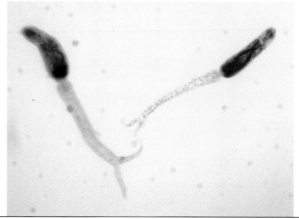

Schistosome ova induce granuloma formation in a variety of organs. While this process is necessary for the transmission of ova, it also leads to the entrapment of numerous ova which never make to the lumen of the appropriate organ for dispersal.

Schistosoma mansoni ova in liver tissue, 400x, c.s.	*Schistosoma mansoni ova* in liver tissue, 1000x, c.s.
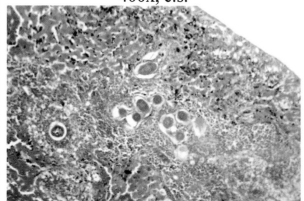	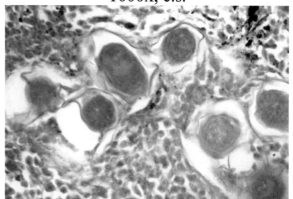

Schistosoma haematobium

Images:

Comparison of ova from 3 species infecting humans:
http://www.k-state.edu/parasitology/625tutorials/Trematodes01.html
Ova of *Schistosoma haematobium*
http://www.k-state.edu/parasitology/625tutorials/Schistosoma02.html
Ova of *Schistosoma japonicum*
http://www.k-state.edu/parasitology/625tutorials/Schistosoma03.html
Ova of *Schistosoma mansoni*
http://www.k-state.edu/parasitology/625tutorials/Trematodes04.html
Schistosoma mansoni mating pair
http://www.k-state.edu/parasitology/625tutorials/Schistosoma01.html

Phylogeny:	Subclass Digenea, Order Strigeata
Preferred definitive host:	Humans
Reservoir hosts:	None
Vector/intermediate host:	Snails-Genus *Bulinus* or Genus *Physopsis*
Geographical location:	Africa and the Middle East
Organs affected:	Adults reside in the venules of the urinary bladder.
Symptoms and Clinical signs:	Initial phase involves abdominal pain, bronchitis, enlargement of the spleen and liver, and diarrhea. Hematuria and pain on urination follow. Because of cellular damage to urinary bladder, malignant tumors may form. Kidneys themselves are sometimes damaged.
Treatment:	Metrifonate, Preziquantel, Niridazole.

Schistosoma mansoni

Images:

Comparison of ova from 3 species infecting humans:
http://www.k-state.edu/parasitology/625tutorials/Trematodes01.html
Ova of *Schistosoma haematobium*
http://www.k-state.edu/parasitology/625tutorials/Schistosoma02.html
Ova of *Schistosoma japonicum*
http://www.k-state.edu/parasitology/625tutorials/Schistosoma03.html
Ova of *Schistosoma mansoni*
http://www.k-state.edu/parasitology/625tutorials/Trematodes04.html
Schistosoma mansoni mating pair
http://www.k-state.edu/parasitology/625tutorials/Schistosoma01.html

Phylogeny:	Subclass Digenea, Order Strigeata
Preferred definitive host:	Humans
reservoir hosts:	Certain monkeys and rodents
Vector/intermediate host:	Snails-Genus *Biomphalaria*
Geographical location:	Northern Africa, Middle East, S. America
Organs affected:	Adults reside in the portal veins of the large intestine
Symptoms and Clinical sings:	Initial phase involves abdominal pain, bronchitis, enlargement of the spleen and liver, and diarrhea. Egg deposition in venules of large intestine induces pseudotubercle formation, resulting in necrosis and ulceration. Cirrhosis and portal hypertension develop as liver becomes damaged. Splenomegaly occurs. Pseudotubercles may develop in the lungs or nervous system.
Treatment:	Oxamniquine, Praziquantel, Niridazole.

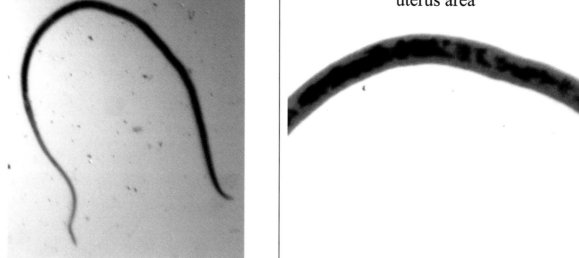

Schistosoma mansoni adult male, 40x

Schistosoma mansoni adult male, closeup of oral sucker, ventral sucker, 100x

Schistosoma mansoni female, 40x

Schistosoma mansoni female, closeup of uterus area

Schistosoma japonicum

Images:

Comparison of ova from 3 species infecting humans:
http://www.k-state.edu/parasitology/625tutorials/Trematodes01.html
Ova of *Schistosoma haematobium*
http://www.k-state.edu/parasitology/625tutorials/Schistosoma02.html
Ova of *Schistosoma japonicum*
http://www.k-state.edu/parasitology/625tutorials/Schistosoma03.html
Ova of *Schistosoma mansoni*
http://www.k-state.edu/parasitology/625tutorials/Trematodes04.html
Schistosoma mansoni mating pair
http://www.k-state.edu/parasitology/625tutorials/Schistosoma01.html

Phylogeny:	Subclass Digenea, Order Strigeata
Preferred definitive host:	Humans
Reservoir hosts:	Rats, dogs, cats, horses, swine, and deer.
Vector/intermediate host:	Snails-Genus *Onchomelania*
Geographical location:	Japan, China, Taiwan, Philippines, Indonesia.
Organs affected:	Adults reside in veins of the small intestine.
Symptoms and Clinical Signs:	Initial phase involves abdominal pain, bronchitis, enlargement of the spleen and liver, and diarrhea. Fibrous nodules containing eggs accumulate on serosal and peritoneal surfaces of the small intestine. Splenomeglay occurs. Cirrhosis and portal hypertension due to live damage follow. Neurological disorders, such as coma or paralysis, may occur due to egg deposition in the brain.
Treatment:	Praziquantel

Class Cestoda

GENERALIZED LIFE CYCLES: CESTODES
(Inner loop = cyclophyllidean pathway)
(Outer loop = pseudophyllidean pathway)

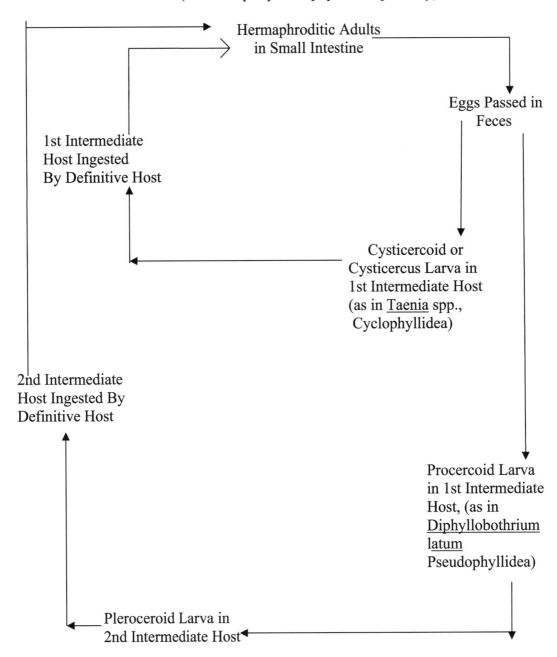

Hermaphroditic Adults
in Small Intestine

Eggs Passed in
Feces

1st Intermediate
Host Ingested
By Definitive Host

Cysticercoid or
Cysticercus Larva in
1st Intermediate Host
(as in Taenia spp.,
Cyclophyllidea)

2nd Intermediate
Host Ingested By
Definitive Host

Procercoid Larva
in 1st Intermediate
Host, (as in
Diphyllobothrium
latum
Pseudophyllidea)

Pleroceroid Larva in
2nd Intermediate Host

Comparison of some pseudophyllidean and cyclophyllidean features

	Pseudophyllidea (e.g. *Diphyllbothrium latum*)	Cyclophyllidea (e.g. *Taenia solium*, *Hymenolepis diminuta*)
Morphology of holdfast organ (scolex)	Adhesive slits 	Adhesive suckers, possibly with a rostellum
# of intermediate hosts	1	2
Morphology of mature proglottids	Vitellaria are always follicular and scattered throughout the proglottid, testes are many 	Have a single compact, postovarian vitelline gland, Number of testes varies from one to several hundred. (*Taeniarhynchus saginata* mature proglottid shown here.)
6 hooks in ovum	Absent	Present

Diphyllbothrium latum (broad tapeworm)

Images:

Egg:
http://www.k-state.edu/parasitology/625tutorials/Platys01.html
Proglottids and eggs:
http://www.k-state.edu/parasitology/625tutorials/Tapeworm05.html
Scolex:
http://www.k-state.edu/parasitology/625tutorials/Tapeworm02.html

Life cycle:

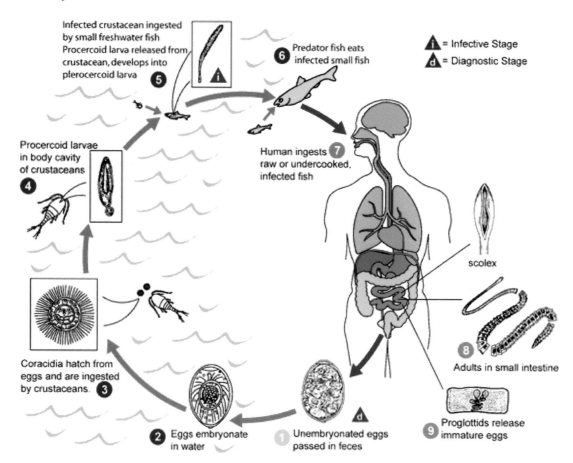

Phylogeny:	Class Cestoda, Order Pseudophyllidea
Preferred definitive host:	Humans
Reservoir hosts:	Piscivorous mammals such as bears
Vector/intermediate host:	1. *Diaptomus* copepods; 2. Fish, particularly

whitefish

Geographical location:	Scandinavia, Russia, Arctic, United States
Organs affected:	Small intestine
Symptoms and clinical signs:	Vague abdominal discomfort. Sometimes pernicious anemia due to vitamin B12 requirement of the parasite. Nausea and diarrhea sometimes occur, but these symptoms are rare.
Treatment:	Niclosamide, Quinacrine, Paromomycin

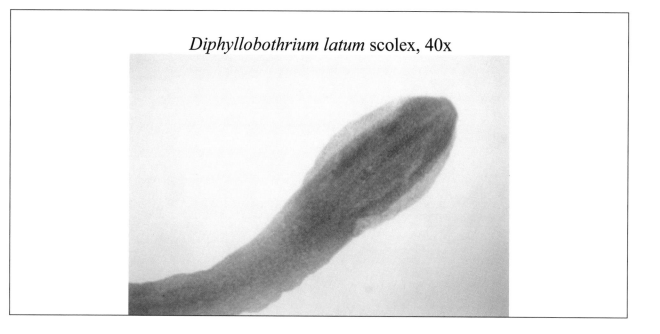

Diphyllobothrium latum scolex, 40x

Diphyllobothrium latum mature proglottid

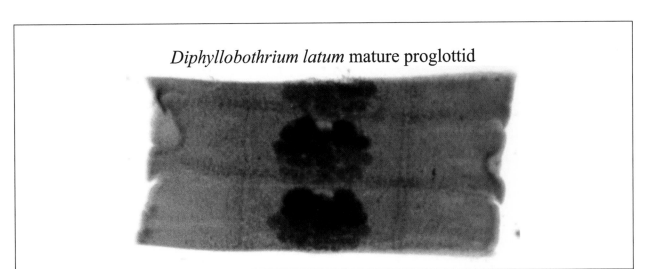

Diphyllobothrium latum gravid proglottid

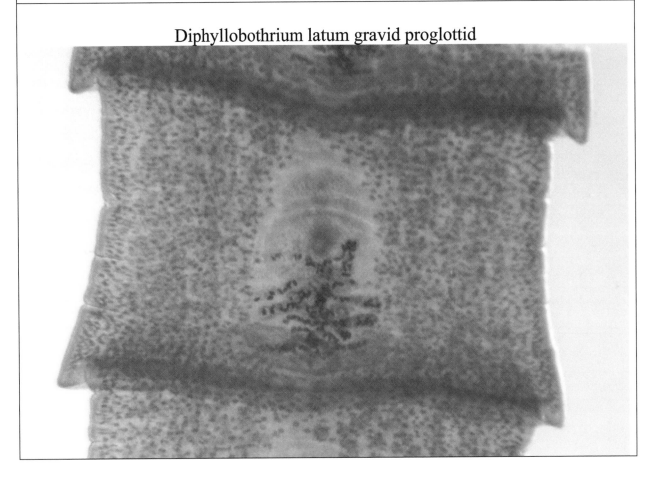

Diphyllobothrium latum ovum,

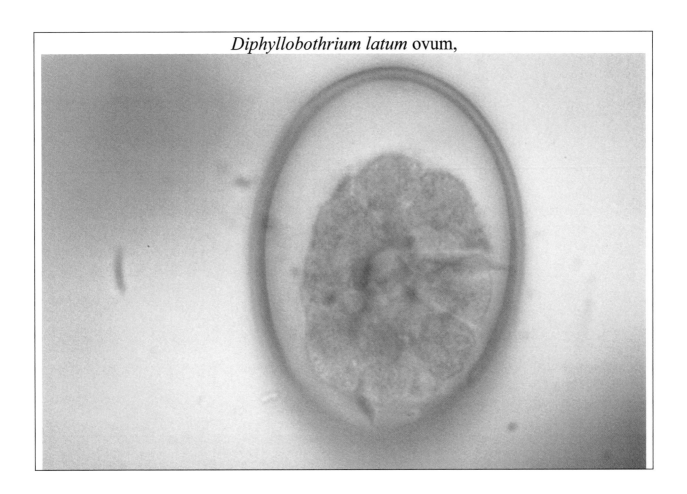

Taenia solium (pork tapeworm)

Images:

Comparison of *Taenia solium* and *Taeniarhynchus saginata* scoleces
http://www.k-state.edu/parasitology/625tutorials/Tapeworm13.html

Phylogeny:	Class Cestoda, Order Cyclophyllidea
Preferred definitive host:	Humans
Reservoir hosts:	None
Vector/intermediate host:	Pigs
Geographical location:	Cosmopolitan
Organs affected:	Adults reside in the small intestine. Cysticerci can reside in heart muscle, brain tissue, or inside the eye.
Symptoms and Clinical signs:	Usually none among adults. Abdominal pain, dizziness, nausea, and diarrhea occur, but are relatively rare. Cysticerci, however, may cause irreparable damage to the eye or heart, may cause necrosis of heart tissue, and may cause severe damage to the central nervous system, leading to epilepsy and hydrocephalus.
Treatment:	For adults, niclosamide, quinacrine, or paromomycin. For cysticerci larvae, surgery is required.

Taenia solium mature proglottid	*Taenia solium* gravid proglottid
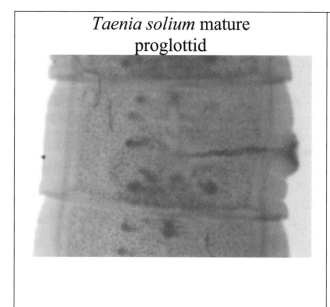	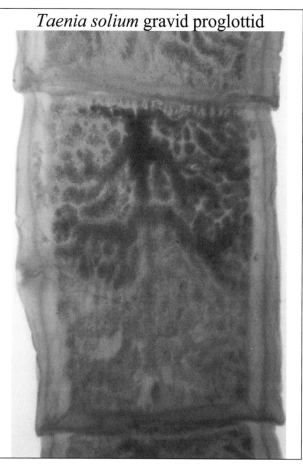

Taenia solium ova

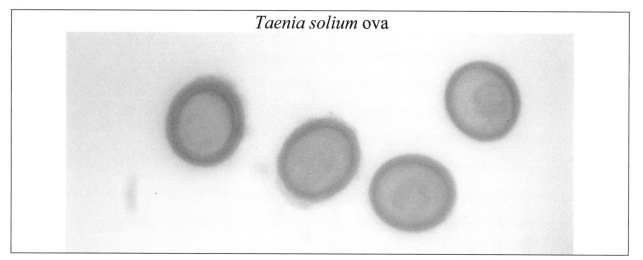

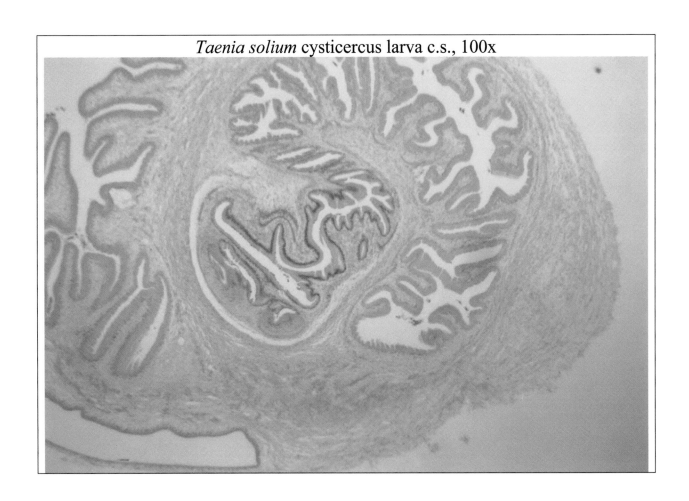

Taenia solium cysticercus larva c.s., 100x

Comparison of *Taenia solium* and *Taeniarhynchus saginata*

	Taenia solium	*Taeniarhynchus saginata*
Mature proglottid		
Gravid proglottid		
Ova (400x mag.)		

Taeniarhynchus saginata (beef tapeworm)

Images:

Comparison of *Taenia solium* and *Taeniarhynchus saginata* scolecesL
http://www.k-state.edu/parasitology/625tutorials/Tapeworm13.html

Phylogeny:	Class Cestoda, Order Cyclophyllidea
Preferred definitive host:	Humans
Reservoir hosts:	None
Vector/intermediate host:	Cattle
Geographical location:	Worldwide, but common in Africa and South America.
Organs affected:	Small intestine
Symptoms and and Clinical Signs:	Usually none, but abdominal pain, headache, diarrhea, and intestinal obstruction may occur in heavy infection.
Treatment:	Niclosamide, quinacrine, paromomycin.

Taeniarhynchus saginata cysticercus larvae

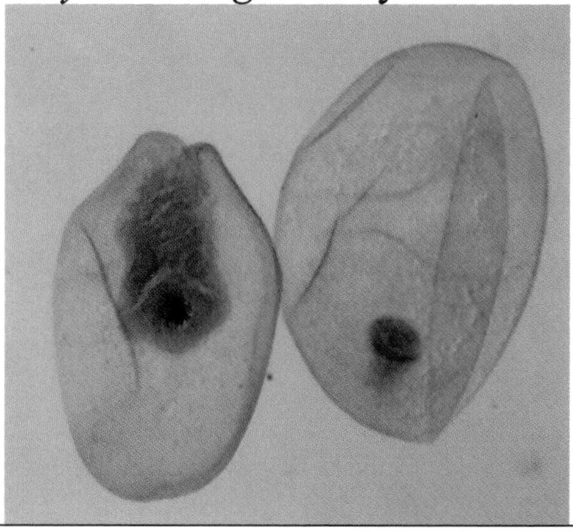

Hymenolepis diminuta (rat tapeworm)

Images:

Ova:
http://www.k-state.edu/parasitology/625tutorials/Hymenolepis.html

Phylogeny:	Class Cestoda, Order Cyclophyllidea
Preferred definitive host:	Rats
Reservoir hosts:	Humans, particularly children.
Vector/intermediate host:	*Tribolium* and/or *Tenebrio* beetles
Geographical location:	Cosmopolitan
Organs affected:	Small intestine
Symptoms and clinical signs:	Usually none
Treatment:	Niclosamide or paromomycin

Hymenolepis diminuta composite
L-R: Scolex and immature proglottids, mature, gravid. 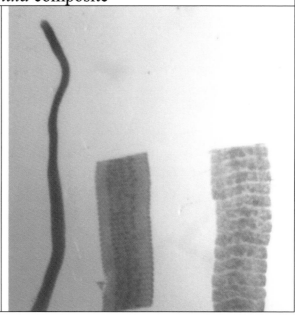

Hymenolepis diminuta scolex

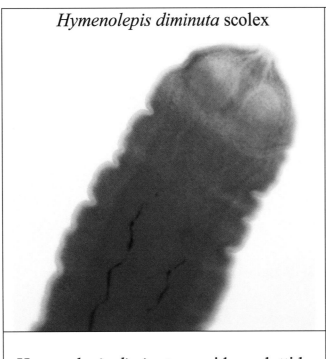

Hymenolepis diminuta gravid proglottids, 40x

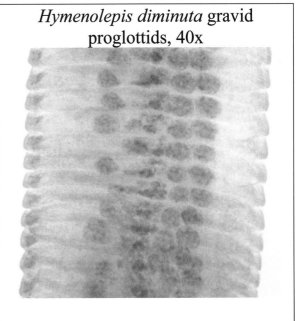

Hymenolepis diminuta gravid proglottids, 100x

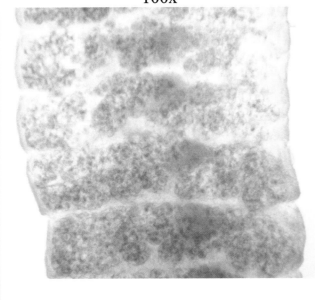

Hymenolepis diminuta cysticercoid larva, 40x

Hymenolepis diminuta cysticercoid larva, 100x	*Hymenolepis diminuta* ovum, 400x

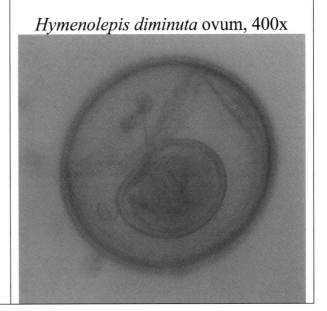

Vampirolepis nana (small rat tapeworm)

Images:

Scolex and ova:
http://www.k-state.edu/parasitology/625tutorials/Vampirolepis.html

Phylogeny:	Class Cestoda, Order Cyclophyllidea
Preferred definitive host:	Rats and mice
Reservoir hosts:	Humans, particularly children.
Vector/intermediate host:	OPTIONAL - Grain beetles, such as *Tribolium* or *Tenebrio*
Geographical location:	Cosmopolitan
Organs affected:	Small intestine
Symptoms and Clinical signs:	Usually none
Treatment:	Niclosamide or paromomycin

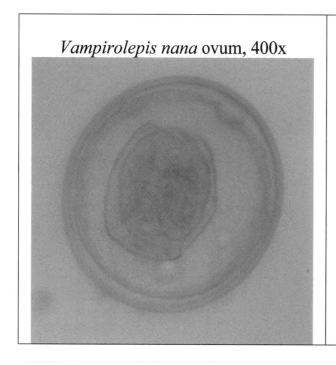
Vampirolepis nana ovum, 400x

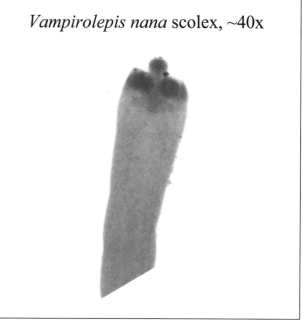
Vampirolepis nana scolex, ~40x

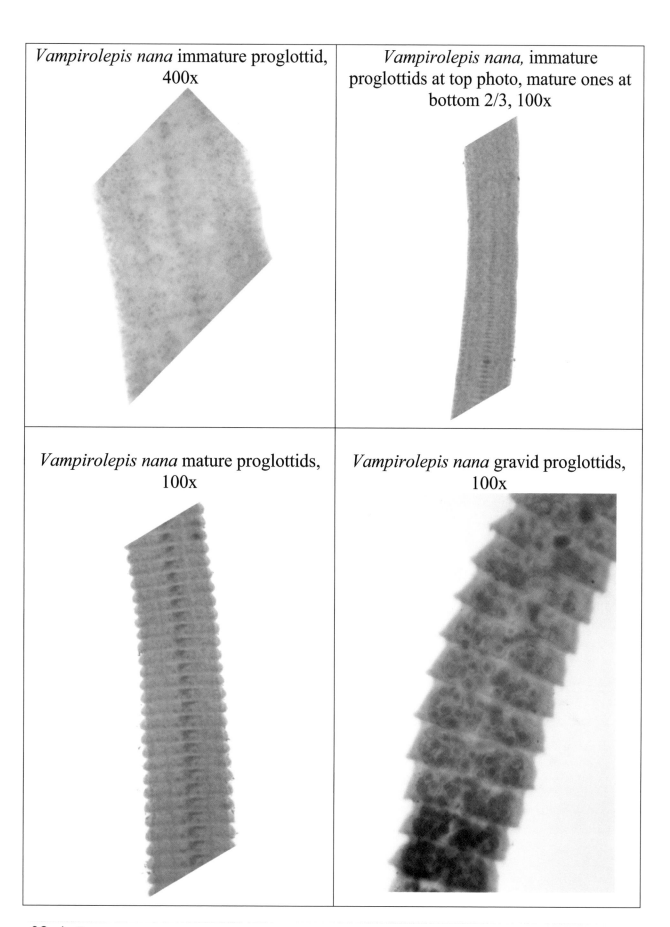

Vampirolepis nana immature proglottid, 400x

Vampirolepis nana, immature proglottids at top photo, mature ones at bottom 2/3, 100x

Vampirolepis nana mature proglottids, 100x

Vampirolepis nana gravid proglottids, 100x

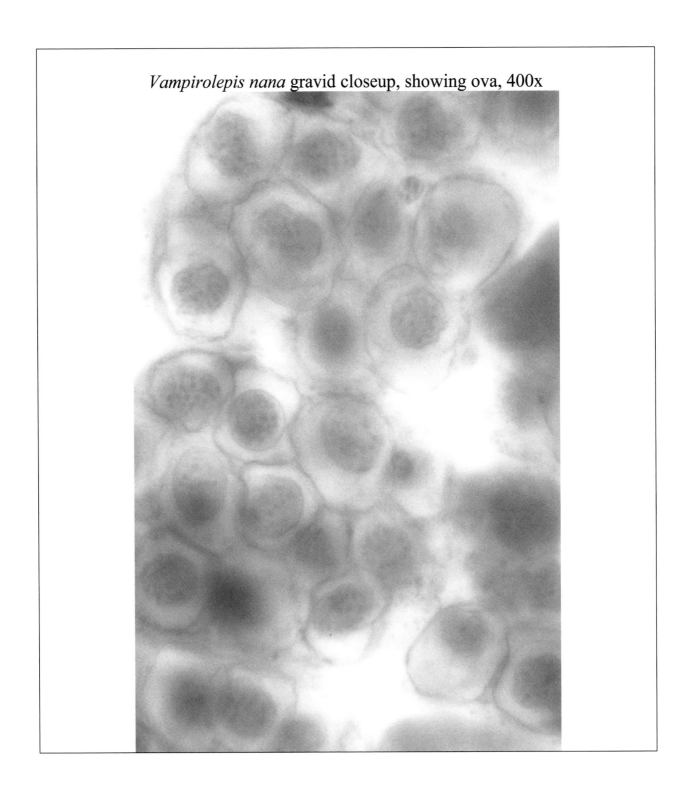

Vampirolepis nana gravid closeup, showing ova, 400x

Dipylidium caninum (dog tapeworm)

Images:

Egg packets:
http://www.k-state.edu/parasitology/625tutorials/Tapeworm06.html
Scolex and mature proglottid:
http://www.k-state.edu/parasitology/625tutorials/Tapeworm01.html

Life cycle:

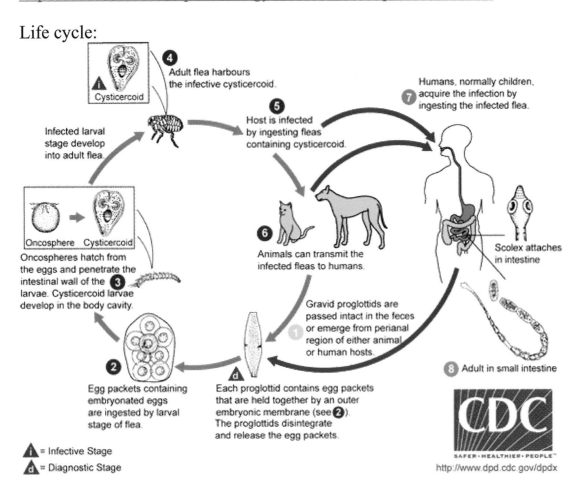

Phylogeny:	Class Cestoda Order Cyclophyllidea
Preferred definitive host:	Domestic dogs and cats
Reservoir hosts:	Humans, particularly children
Vector/intermediate host:	Fleas
Geographical location:	Cosmopolitan

Organs affected: Small intestine

Symptoms and
    Clinical signs: Usually none, although abdominal pain, headache, diarrhea, and verminous intoxication may occur, particularly in patients with heavy infection.

Treatment: Niclosamide, quinacrine, or praziquantel.

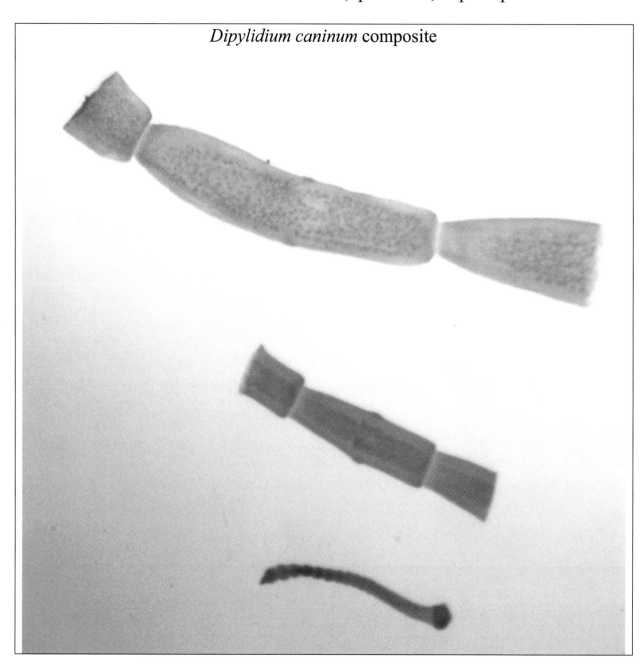

Dipylidium caninum composite

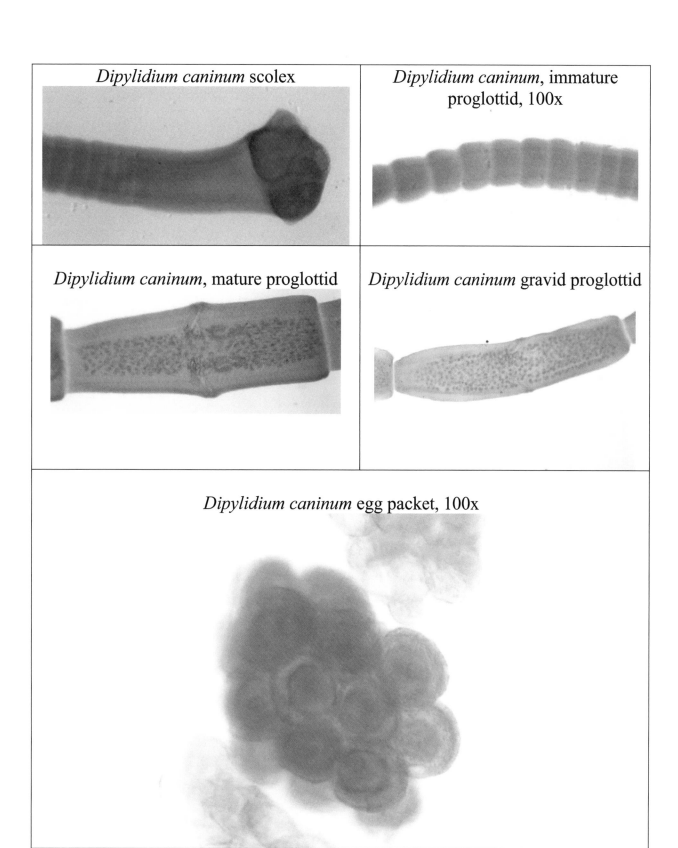

Dipylidium caninum scolex

Dipylidium caninum, immature proglottid, 100x

Dipylidium caninum, mature proglottid

Dipylidium caninum gravid proglottid

Dipylidium caninum egg packet, 100x

Echinococcus granulosus

Images:

Entire worm, scolex, protoscoleces:
http://www.k-state.edu/parasitology/625tutorials/Tapeworm03.html

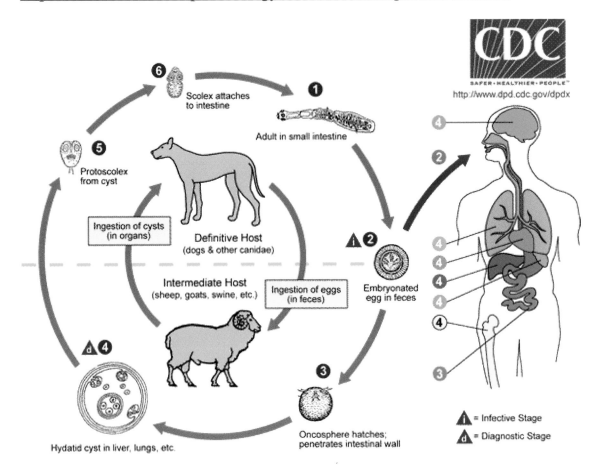

Phylogeny:	Class Cestoda, Order Cyclophyllidea
Preferred definitive host:	Carnivores, particularly dogs
Reservoir hosts:	Other mammals
Vector/intermediate host:	Herbivores, particularly moose, reindeer, goats, camels and sheep. Humans are 'accidental' hosts.
Geographical location:	Cosmopolitan

Organs affected:	Cysts may develop in bone marrow, nervous system, liver, and lungs.
Symptoms and Clinical signs:	Specific symptoms depend on the site of cyst formation. In general, the presence of the cyst will induce pressure in organs and cause necrosis. Hydatid fluid can induce anaphylactic shock and death.
Treatment:	Surgical removal of the cyst is required.

Echinococcus granulosus entire worm

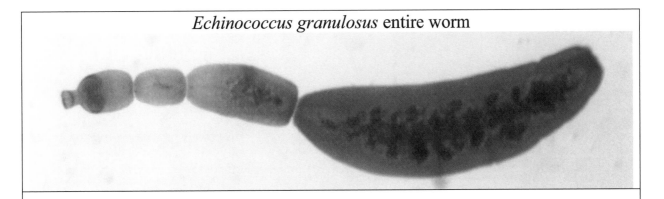

Echinococcus granulosus anterior portion of the worm, showing scolex, immature & mature proglottids

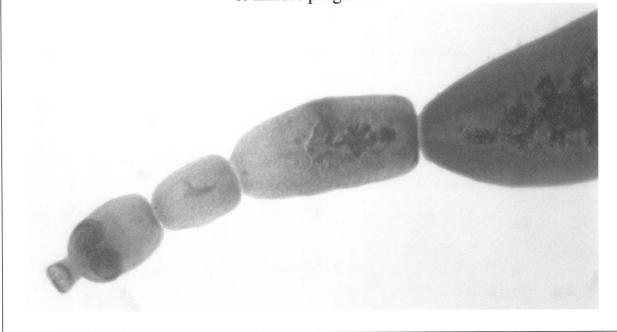

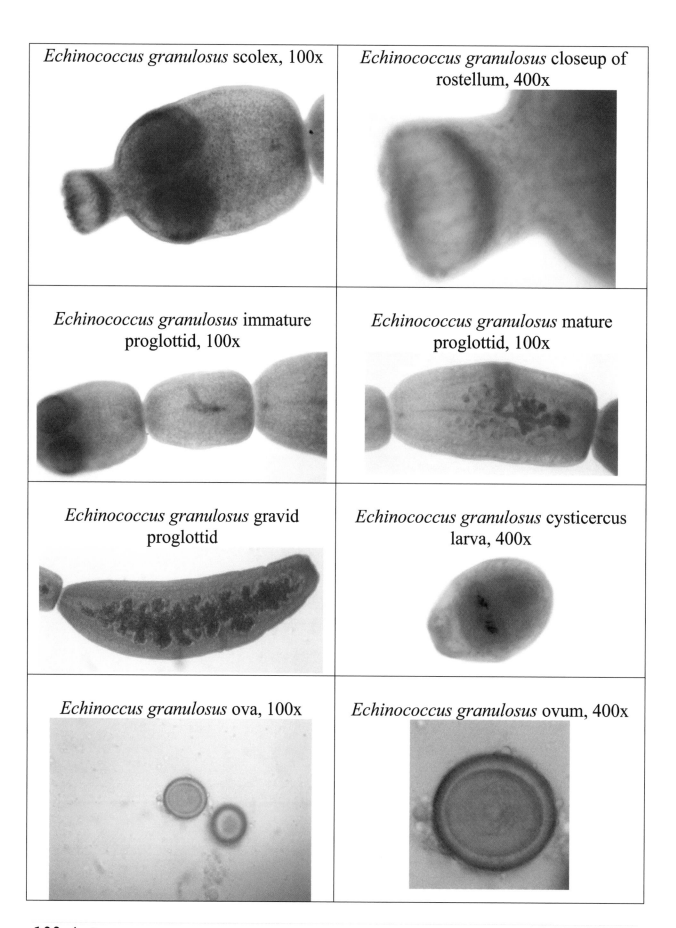

Echinococcus granulosus scolex, 100x

Echinococcus granulosus closeup of rostellum, 400x

Echinococcus granulosus immature proglottid, 100x

Echinococcus granulosus mature proglottid, 100x

Echinococcus granulosus gravid proglottid

Echinococcus granulosus cysticercus larva, 400x

Echinoccus granulosus ova, 100x

Echinococcus granulosus ovum, 400x

Echinococcus granulosus, hydatid cyst c.s., 100x	*Echinococcus granulosus* hydatid sand w.m., 100x
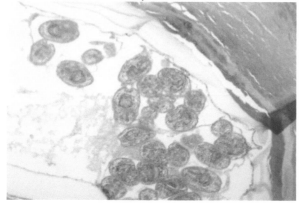	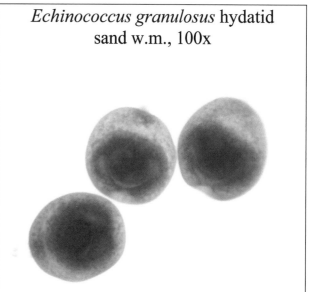

PHYLUM NEMATODA

- Triploblastic;
- Dioecious;
- Pseudocoelomate, i.e. digestive system is not bounded by mesodermally-derived tissue (see cross-section of adult *Ascaris lumbricoides* below;

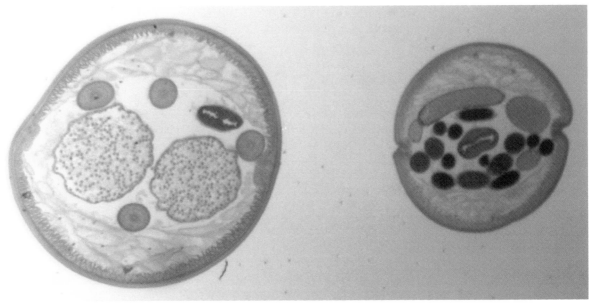

- Tough cuticle used for taxonomic purposes, already makes nematodes relatively resistant to drug treatments;
- Four-molting events leading to sexual maturity; hormonal control similar to that of arthropods;
- Digestive system complete.

GENERALIZED LIFE CYCLES: INTESTINAL NEMATODES

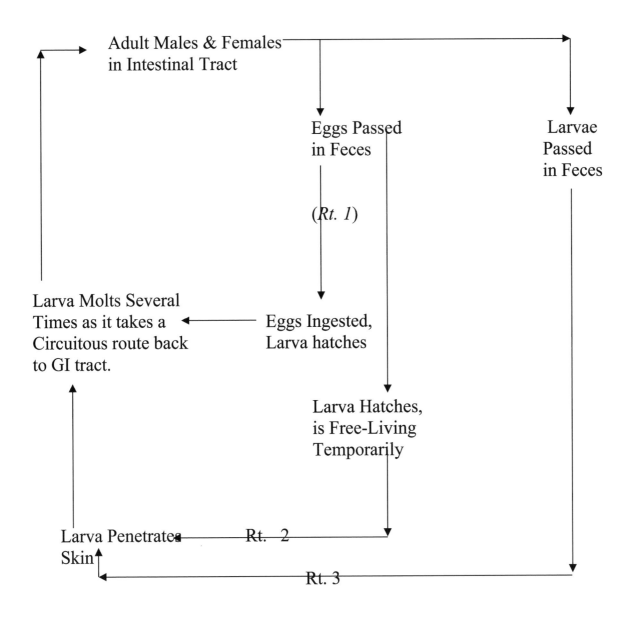

#1: Ascaris
#2: Hookworms
#3: Strongyloides

Trichuris trichiura (whipworm)

Images:

Ova of *Trichuris vulpis:*
http://www.k-state.edu/parasitology/625tutorials/Nematodes02.html

Phylogeny	Order Trichurata.
Preferred Definitive host:	Humans
Reservoir hosts:	Reported in monkeys and dogs.
Vector/intermediate host:	None
Geographical location:	Cosmopolitan, but most frequent in tropical countries
Organs affected:	Human cecum, appendix, ileum.
Symptoms and Clinical signs:	In heavy infections, patients show small blood-streaked diarrheal stools, abdominal pain and tenderness, nausea and vomiting, anemia and weight loss. Prolapse of the rectum has occurred in some cases.
Treatment:	Mebendazole.

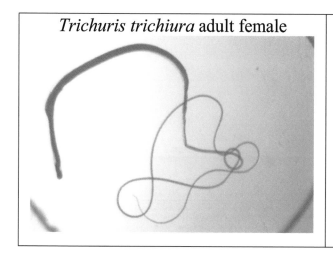

Trichuris trichiura adult female

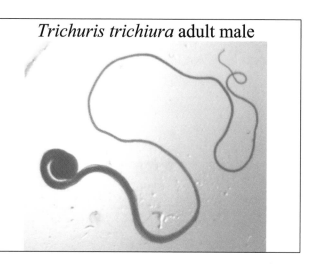

Trichuris trichiura adult male

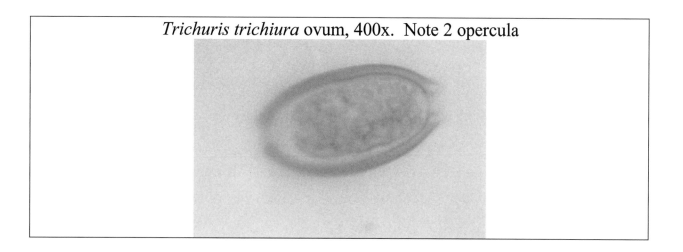

Trichuris trichiura ovum, 400x. Note 2 opercula

Trichinella spiralis

Images:

```
Adult female:
```
http://www.k-state.edu/parasitology/625tutorials/Trichinella01.html
```
Larvae in muscle tissue:
```
http://www.k-state.edu/parasitology/625tutorials/Trichinella02.html
```
World distribution maps of species in genus:
```
http://www.k-state.edu/parasitology/625tutorials/Trichinella03.html

```
Life cycle:
```

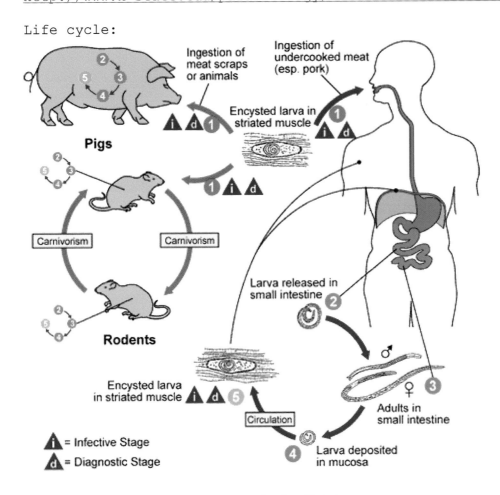

Phylogeny: Order Trichurata

Preferred definitive host: Humans

Reservoir hosts: Carnivorous mammals, including rodents and pigs.

Vector/intermediate host: None

| Geographical location: | Cosmopolitan, but most frequently found in circumboreal countries. |

Geographical location: Cosmopolitan, but most frequently found in
                          circumboreal countries.

Organs affected: 1. Initial phase - Mucosa of small intestine
                          2. Penetration phase - larvae lodging in
                          striated muscle, myocardium, brain
                          and meninges.

Symptoms and
  Clinical signs: 1. Initial phase - Nausea, vomiting,
                          diarrhea, headache
                          2. Penetration phase - Edema, conjunctivitis,
                     fever, chills, dyspnea, muscle paint. Other symptoms
                          include EKG disorders, headache, mental apathy,
                          delirium, coma.

Treatment: Symptoms are relieved with analgesics and
                          corticosteroids. Thiabendazole is effective among
                          experimental animals.

Trichinella spiralis 3rd stage larvae in muscle tissue, 100x	*Trichinella spiralis* larva in muscle tissue, 400x
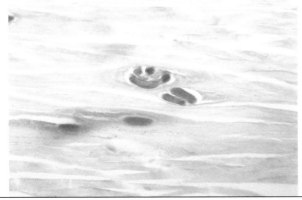	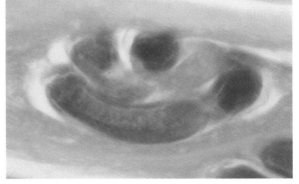
Trichinella spiralis adult female, 40x	*Trichinella spiralis* adult male, 40x

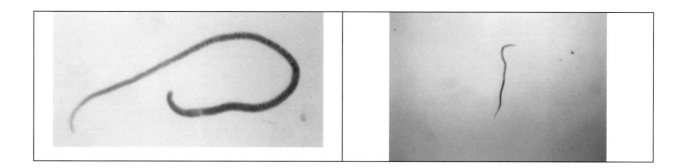

Strongyloides stercoralis

Images:
L1 larva:
http://www.k-state.edu/parasitology/625tutorials/Miscellaneous02.html

Preferred definitive host:	Humans
Reservoir hosts:	Other primates, dogs, cats, other mammals
Vector/intermediate host:	None
Geographical location:	Cosmopolitan
Organs affected:	Adult worms are generally found in the small intestine. Occasionally, they will also be found the respiratory, biliary, or pancreatic system.
Symptoms and Clinic signs:	I. Invasion phase - Penetration of the skin by larvae will cause slight hemorrhage and swelling. The site of entry will show intense itching. Worms may also enter the body orally by ingestion with contaminated food. Worms which follow the oral route bypass the pulmonary phase.
	II. Pulmonary phase - Damage to lung tissue causes massive-host cell reactions. Symptoms include a burning sensation in the chest, nonproductive cough, bronchial pneumonia.
	III. Intestinal phase - Among immunocompetent individuals, the infection is generally asymptomatic. Among immunosuppressed individuals, the problems arising from hyperinfection can become life-threatening. There is persistent diarrhea, and migrating worms are known to transport coliform bacteria throughout the body, and thereby may cause a gram-negative encephalitis by entry into the nervous system.

Treatment: Thiabendazole, Cambendazole.

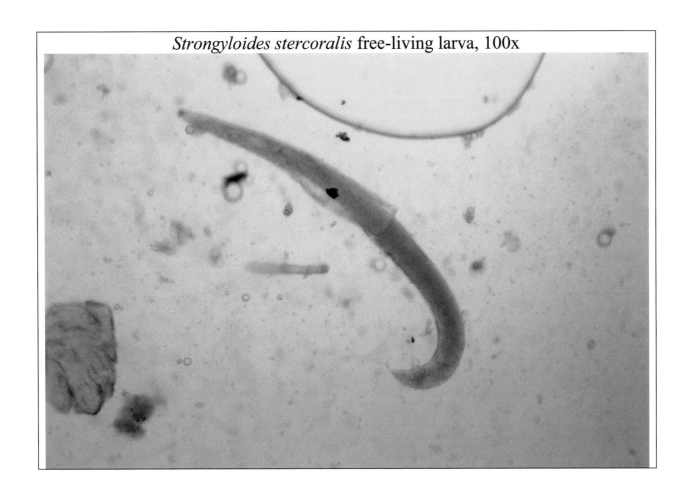

Strongyloides stercoralis free-living larva, 100x

Ancylostoma caninum (dog hookworm, causes dermal larva migrans among humans) – serves as a suitable model for hookworms

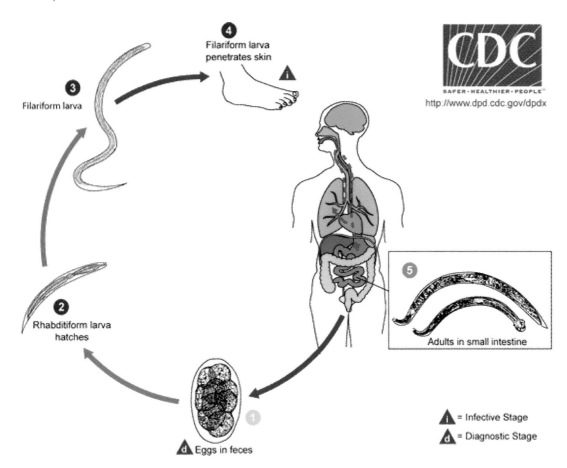

Images:

Ancylostoma caninum adults:
http://www.k-state.edu/parasitology/625tutorials/Ancylostoma.html
Necator americanus buccal armature and ova:
http://www.k-state.edu/parasitology/625tutorials/Nematodes09.html

Phylogeny: Phylum Nematoda, Order Strongylata

Preferred definitive host: Domestic dogs and cats

Reservoir hosts: None. Humans are 'accidental' hosts

Vector/intermediate hosts: None

Geographical location: Northern Hemisphere

Organs affected: Skin

Symptoms and
    Clinical Signs: Creeping eruption, characterized by inflammation
                              and itching along migration pathways of larvae

Treatment: Thiabendazole ointment.

Ancylostoma caninum adult female and male, for comparison

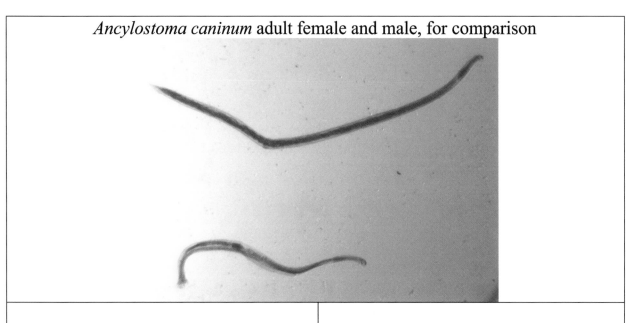

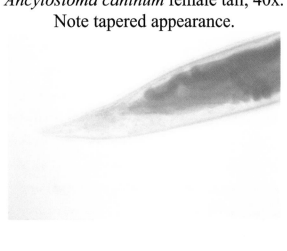

Ancylostoma caninum female tail, 40x.
Note tapered appearance.

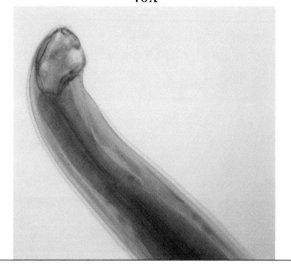

Ancylostoma caninum female, oral area,
40x

Ancylostoma caninum male tail, 40x. Note copulatory bursa.

Ancylostoma caninum male oral area, 40x.

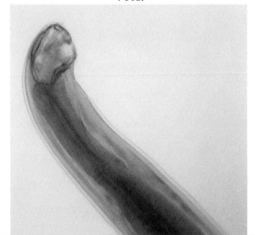

Ancylostoma duodenale (Old World hookworm)

Images:

Ancylostoma caninum adults:
http://www.k-state.edu/parasitology/625tutorials/Ancylostoma.html
Necator americanus buccal armature and ova:
http://www.k-state.edu/parasitology/625tutorials/Nematodes09.html

Life cycle:

Phylogeny:	Phylum Nematoda, Order Strongylata
Preferred definitive host:	Humans
Reservoir hosts:	None
Vector/intermediate host:	None
Geographical location:	Cosmopolitan
Organs affected:	Larvae affect the skin and lungs, while the adults affect the small intestine.
Symptoms and Clinical signs:	1. Cutaneous phase: Itching of skin 2. Pulmonary phase: Bronchitis, pneumonitis 3. Intestinal phase: None in light infection. In heavy heavy infections, anemia leading to dyspnea on exertion, weakness and dizziness occur. Heart shows atrophy, and children may show physical, mental, or sexual retardation.
Treatment:	Mebendazole, pyrantel pamoate and supplemental iron to offset anemia.

Necator americanus (New World hookworm)

Images:

Ancylostoma caninum adults:
http://www.k-state.edu/parasitology/625tutorials/Ancylostoma.html
Necator americanus buccal armature and ova:
http://www.k-state.edu/parasitology/625tutorials/Nematodes09.html

Phylogeny:	Phylum Nematoda, Order Strongylata
Preferred definitive host:	Humans
Reservoir hosts:	None
Vector/intermediate host:	None
Geographical location:	Cosmopolitan
Organs Affected:	Larvae affect the skin and lungs, while the adults affect the small intestine
Symptoms and Clinical Signs:	1. Cutaneous phase: Itching of skin 2. Pulmonary phase: Bronchitis, pneumonitis. 3. Intestinal phase: None in light infection. In heavy infections, anemia leading to dyspnea on exertion, weakness and dizziness occur. Heart shows atrophy, and children may show physical, mental or sexual retardation.
Treatment:	Mebendazole, pyrantel pamoate and supplemental iron to offset anemia.

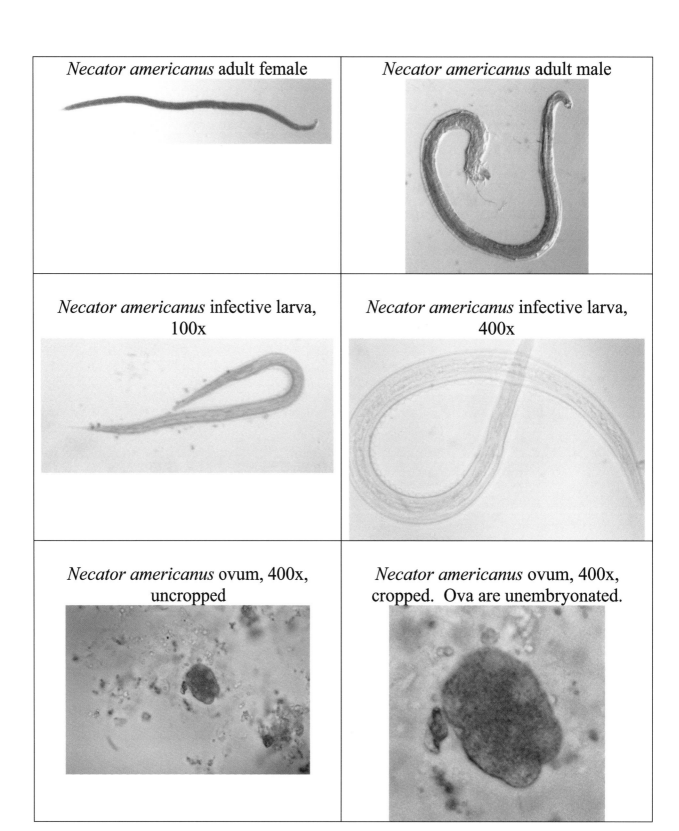

Necator americanus adult female

Necator americanus adult male

Necator americanus infective larva, 100x

Necator americanus infective larva, 400x

Necator americanus ovum, 400x, uncropped

Necator americanus ovum, 400x, cropped. Ova are unembryonated.

COMPARISON OF HUMAN HOOKWORMS

	ANCYLOSTOMA DUODENALE	NECATOR AMERICANUS
MOUTH PARTS	TWO VENTRAL PLATES WITH TEETH	VENTRAL & DORSAL CUTTING PLATES
LENGTH OF MALE (INCHES)	8-11	7-9
LENGTH OF FEMALE (INCHES)	10-13	9-11
AVERAGE FEEDING RATE (ML BLOOD/DAY)	.15	.03
DAILY EGG PRODUCTION	25,000 to 30,000	9,000

CORRELATION BETWEEN NECATOR WORM

BURDEN AND PATIENT STATUS

WORM BURDEN	LEVEL OF PATHOLOGY
< 25	SYMPTOMLESS
25 TO 100	LIGHT SYMPTOMS
100 TO 500	MODERATE
500 TO 1000	SEVERE WITH GRAVE DAMAGE
> 1000	POSSIBLY FATAL

ANCYLOSTOMA, TAKING MORE BLOOD PER DAY, WILL REQUIRE FEWER WORMS TO GENERATE SEVERE PATHOLOGY.

CHRONIC INFECTIONS LEAD TO MENTAL DULLNESS, PHYSICAL RETARDATION, HEART FAILURE, DEATH.

Ascaris lumbricoides (large roundworm)

Images:

Ova:
http://www.k-state.edu/parasitology/625tutorials/Ascaris01.html

Parasites, like politics, make strange bedfellows:
http://facstaff.cbu.edu/~seisen/worm.jpg

Life cycle:

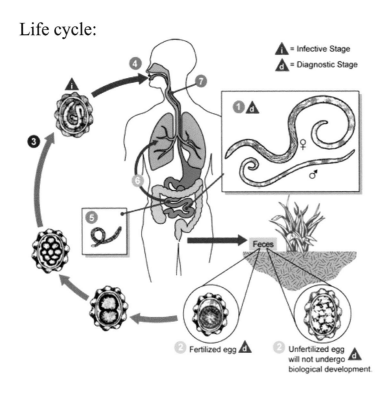

Phylogeny:	Phylum Nematoda, Order Ascaridata
Preferred definitive host:	Humans
Reservoir hosts:	None, but Ascaris suum in swine is very similar
Vector/intermediate host:	None
Geographical location:	Cosmopolitan

Organs affected:	Adults reside in the lumen of the small intestine

Symptoms and
Clinical Signs:

1. Initial phase: Vague symptoms arising from inflammatory responses
2. Pulmonary phase: Edema, pneumonitis
3. Intestinal phase: Abdominal pain, asthma, insomnia. Use of ineffective drugs will stimulate migration, leading to serious and sometimes fatal results. Worms have been known to escape through the nares, and to penetrate the intestinal wall and emerge from the body wall. Worms will also invade visceral organs.

Treatment: Piperazine citrate, mebendazole, pyrantel pamoate.

Ascaris lumbricoides adult female and male c.s.

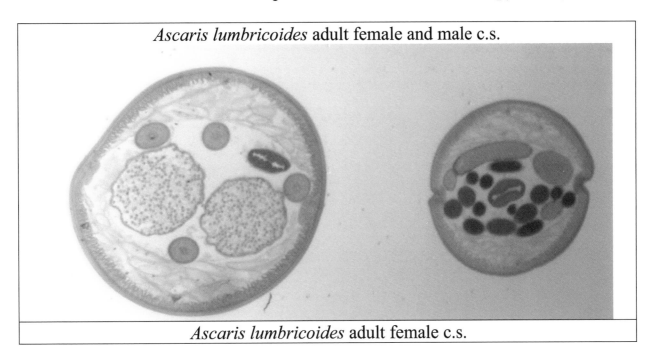

Ascaris lumbricoides adult female c.s.

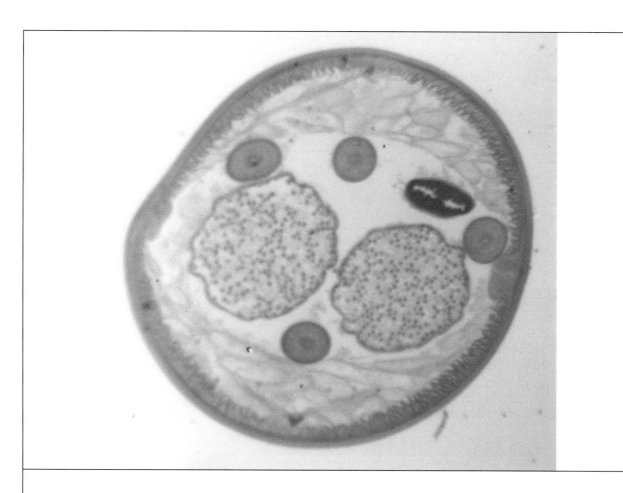

Ascaris lumbricoides adult male c.s.

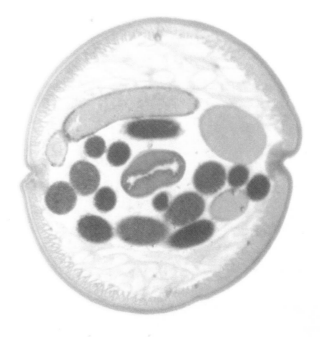

Ascaris lumbricoides ovum, 400x, cropped. Note "mammilated" egg surface.

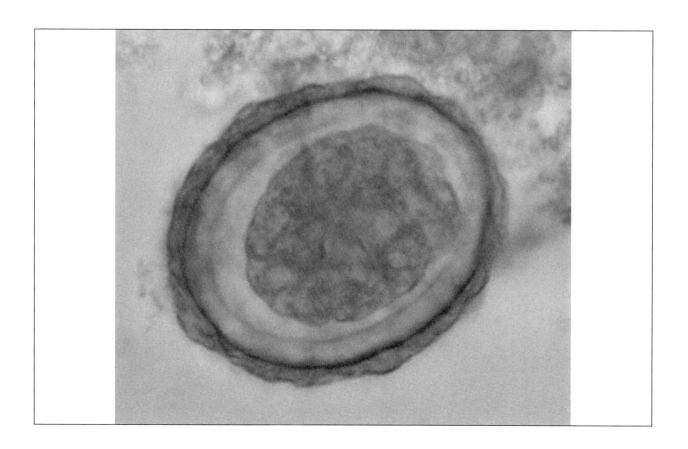

Toxocara cati and *T. caninum* (visceral larva migrans)

Images:

Toxocara cati ova:
http://www.k-state.edu/parasitology/625tutorials/Nematodes03.html

Life cycle:

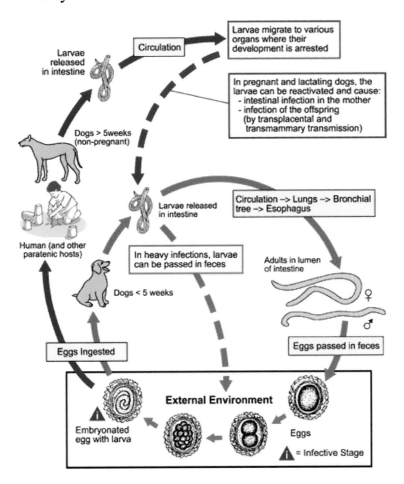

Phylogeny: Phylum Nematoda, Order Ascaridata

Preferred definitive host: Cats and Dogs

Reservoir hosts: None. Humans are 'accidental' hosts

Vector/intermediate host: None

Geographical location: Cosmopolitan

Organs affected:	Liver, lungs, eye, brain, cardiac muscle, kidney
Symptoms and Clinical Signs:	Visceral larva migrans results in inflammation and eosinophilic granulomas in organs. Pneumonitis, hepatomegaly, spleen enlargement, anemia, iritis and hemorrhage of the eye are common symptoms.
Treatment:	Thiabendazole

Toxocara canis ovum, 400x

Enterobius vermicularis (pinworm)

Images:
Ova:
http://www.k-state.edu/parasitology/625tutorials/Nematodes08.html

Phylogeny: Phylum Nematoda, Order Oxyurata

Preferred definitive host: Humans

Reservoir hosts: None

Vector/intermediate host: None

Geographical location: Temperate zone, especially in Europe and North America

Organs affected: Ileocecal region of the intestine

Symptoms and
clinical signs: Generally asymptomatic, but heavy infections will result
                        in disturbed sleep. This, in turn, will debilitate the
                        patient. Itching and pruritis are observed. Minute
                        ulcerations of the intestinal mucosa and fatal
                        subserosal penetration has been reported.

Treatment: Piperazine citrate, pyrvinium pamoate, mebendazole.

This might be a little intense to watch...

http://blog.sciam.com/index.php?p=166&more=1&c=1&tb=1&pb=1#more166

"Do not look at this....
unless you are fascinated by all things parasitological..."

A video clip that accompanies an article in New England Journal
of Medicine
http://content.nejm.org/cgi/content/full/354/13/e12

"A colonoscopy was ordered and revealed multiple mobile 1-cm
worms, *Enterobius vermicularis,* in the cecum "

Enterobius vermicularis ovum, 400x. Note asymmetric shape and clear space between the embryo and egg shell.

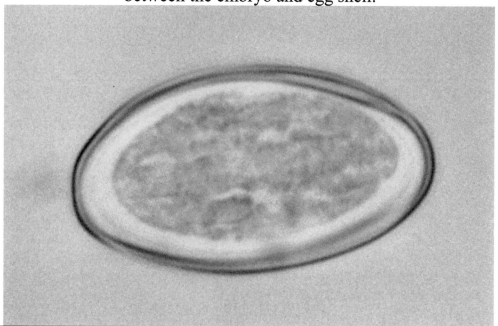

Enterobius vermicularis female. Note long tapered tail.

Enterobius vermicularis male.

LIFE CYCLE OF FILARIAL NEMATODES

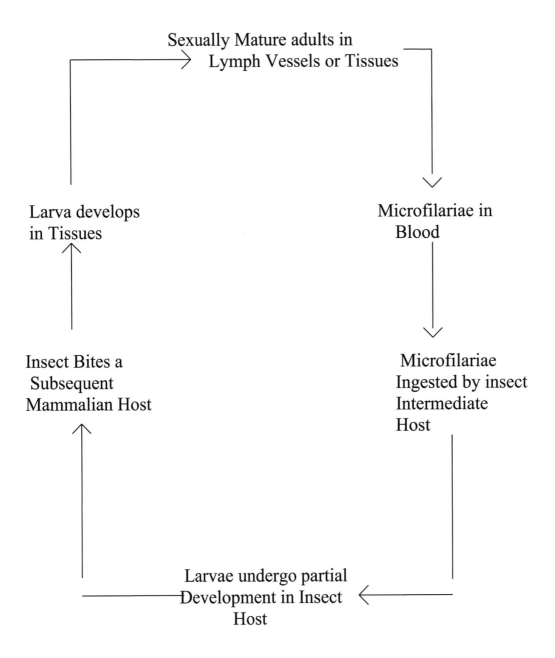

Wuchereria bancrofti (elephantiasis)

An e-article on filariasis:
http://www.emedicine.com/MED/topic794.htm

Images:

Microfilariae in blood:
http://www.k-state.edu/parasitology/625tutorials/Nematodes07.html

Maybe you shouldn't go here:
Elephantiasis mammaria
http://www.cbu.edu/~seisen/mammaryeleph.jpg

Maybe you shouldn't go here either:
Scrotal elephantiasis
http://vatican.rotten.com/testicle/eleph-row.html
http://elephantiasis.freeyellow.com/elephantiasis_clip.mpg

Biology as art, or maybe it's art as biology:
http://www.cbu.edu/~seisen/doorknocker.jpg

Life cycle:

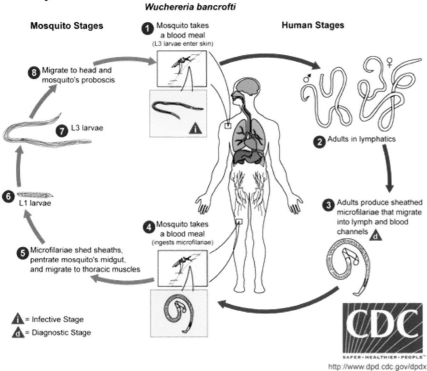

Phylogeny: Phylum Nematoda, Order Filariata

Preferred definitive host: Humans

Reservoir hosts: None

Vector/intermediate host: Mosquitos, including the genera *Anopheles, Aedes, Culex,* and *Mansonia*

Geographical location: Central Africa, Turkey, India, Southeast Asia, Australia, South Pacific Islands

Organs affected: Microfilariae are in the blood, and adults reside in major lymphatic ducts.

Symptoms and
    Clinical signs: Microfilariae are virtually symptomless. Inflammation caused by the presence of adults leads to chills, fever, and toxemia. Lymph vessels become swollen, leading to swelling of organs and the accumulation of lymph in urine.

Treatment: Microfilariae and adults are killed by diethylcarbamazine. Mechanical damage is treated with pressure bandages or surgical removal of elephantoid tissue.

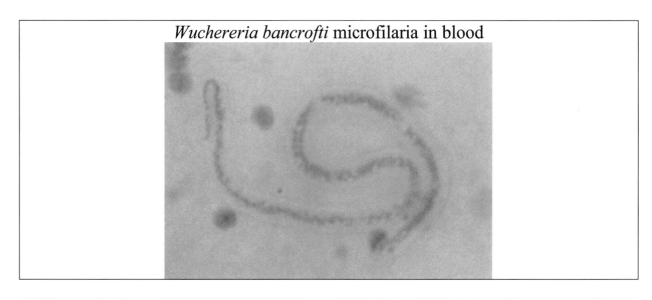
Wuchereria bancrofti microfilaria in blood

Onchocercus volvulus (river blindness)

Images:
Section of skin nodule:

Phylogeny: Phylum Nematoda, Order Filariata

Preferred definitive host: Humans

Reservoir hosts: None

Vector/intermediate host: *Simulium* blackflies

Geographical location: Africa, South and Central America

Organs affected: Microfilariae are in skin, while adults
                 reside in subcutaneous tissue.

Symptoms and
     Clinical signs: Adults cause relatively minor problems. their presence
                           lead to formation of subcutaneous nodules, but
                           they are relatively benign and elicit no pain.
                           Microfilariae cause dermatitis, skin lesions,
                           depletion of vitamin A, and blindness due to
                           corneal invasion.

Treatment: Surgical removal of nodules, and administration of
                    diethylcarbamazine and/or suramin.

Onchocerca volvulus microfilaria from smear of tumor, 400x	*Onchocerca volvulus* c.s. of tumor, 400x
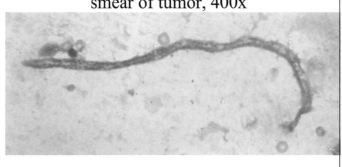	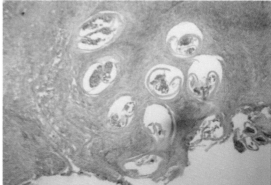

Dirofilaria immitis (heartworm)

Images:

Adults extracted from dog heart:
http://www.k-state.edu/parasitology/625tutorials/Nematodes06.html

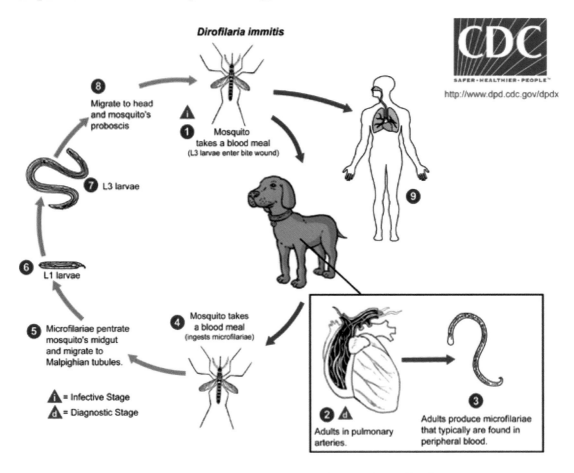

Phylogeny: Phylum Nematoda, Order Filariata

Preferred definitive host: Dogs

Reservoir hosts: None. Humans are rare 'accidental' hosts. Cats are rarely
                                      infected.

Vector/immediate host: Anopheline mosquitos

Geographical location: Cosmopolitan

Organs affected: Right heart and pulmonary artery

Symptoms and
    Clinical signs: Microfilariae are virtually symptomless. However, the
                         presence of a large number of adults will cause
                         right heart failure and pulmonary complications in
                         dogs. In humans, the symptoms are vague and
                         unpredictable. It takes significantly less worms to
                         elicit such symptoms in cats.

Treatment: Adults are destroyed by thiacetarsamide sodium solution,
                         and microfilariae are destroyed by the oral
                         administration of dithiazamine iodide. Heartworm
                         disease is prevented with diethylcarbamazine.

Dirofilaria immitis w.m. from prepared slide, 400x	*Dirofilaria immitis* w.m. in thick smear prepared at CBU, 400x
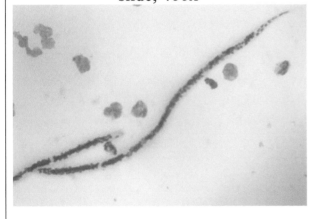	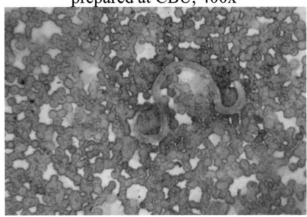

Dracunculus medinensis (Guinea worm)

Images:

 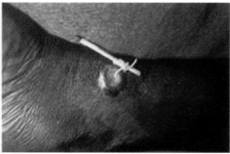

Extracting a Guinea worm from the ankle by wrapping it around a stick. Photo credit: WHO Collaborating Center at CDC archives.

Phylogeny: Phylum Nematoda, Order Camallanata

Preferred definitive host: Humans

Reservoir hosts: Other mammals

Vector/intermediate host: Copepods, genus *Cyclops*

Geographical location: Africa, Southwestern Asia, Northeastern South America, West Indies.

Organs affected: Adults reside in subcutaneous tissues, especially in the legs, arms, shoulder and trunk.

Symptoms and
   Clinical signs: Blister formation is accompanied by urticaria, erythema, dyspnea, vomiting, pruritus, all of which are of allergic nature. Severe inflammation may occur if worm is snapped.

Treatment: Metronidazole, niridazole, thiabendazole.

PHYLUM ANNELIDA (CLASS HIRUDINEA)
CHARACTERISTICS OF MAJOR ANIMAL PHYLA

PHYLUM   ANNELIDA

SYSTEM	TYPE/REMARKS
1. CIRCULATORY	Some oligochaetes have an open system. Most have a closed system. Earthworms have five pairs of contractile esophageal vessels = hearts. Hemoglobin present
2. RESPIRATORY	Most show diffusion through the body wall. Some polychaetes have parapodia or gills.
3. EXCRETORY	Protonephridial or metanephridial type. Aquatic groups excrete ammonia. Earthworms secrete urea.
4. DIGESTIVE	Complete
5. SKELETAL	Generally Hydrostatic
6. NERVOUS	Cerebral ganglia, two closely fused ventral nerve cords, various other ganglia.
7. TYPE OF COELOM	True
8. MUSCULAR	Circular, longitudinal, oblique
9. REPRODUCTIVE	Polychaetes are dioecious, oligochaetes and leeches are hermaphroditic

MAJOR GROUPS OF ANNELIDS

CLASS POLYCHAETA

CLASS OLIGOCHAETA (INCLUDES EARTHWORMS)

CLASS HIRUDINEA: LEECHES

It conjures up SUCH a lovely image:

From the April 1854 issue of Scientific American:

A Lovely Place -
"Dr. Hooker, in his 'Himalayan Journals,' just published, gives the following sketch of a pleasant excursion on the Nepaulese Himalaya: 'Leeches swarmed in incredible profusion in the streams and damp grass, and among the bushes; they got into my hair, hung on my eyelids, and crawled up my legs and down by my back. I repeatedly took upwards of a hundred from my legs where they collected in clusters on the instep; the sores which they produced were not healed for five months, and I retain the scars to the present day.'"

:)

Limnatis SPP. (AQUATIC LEECHES)

Phylogeny: Phylum Annelida, Class Hirudinea

Preferred definitive host: Humans

Reservoir hosts: None

Vector/intermediate host: None

Geographical location: Far East

Organs affected: Primarily skin, but the worm is small enough that the respiratory and digestive systems will be affected. Bathers frequently have infections in the vagina, urethra or eyes.

Symptoms and clinical signs: Pain, hemorrhage

Treatment: Removal of worms

Haemadipsa SPP. (terrestrial leech)

Phylogeny: Phylum Annelida, Class Hirudinea

Preferred definitive host: Humans

Reservoir hosts: None

Vector/intermediate host: None

Geographical location: Far East

Organs affected: Skin

Symptoms and clinical signs: Painless, unnoticed wounds on
                     skin take a considerable time to
                     clot due to anticoagulant used by
                     leech to suck its blood meal.

Treatment: Removal of worm by use of either a local anesthetic, a
                     strong salt solution, or a lighted match. Repellents
                     such as dimethyl phthalate are used on clothing.

Hirudo medicinalis (Medicinal leech)

Phylogeny   Phylum Annelida, Class Hirudinea

Preferred definitive host: Humans

Reservoir hosts: None

Vector/intermediate host: None

Geographical location: Cosmopolitan, but they are cultivated in numerous areas

Organs affected: Skin

Symptoms and clinical
signs: Painless, unnoticed wounds on skin take a considerable
                         length of time to clot due to anticoagulant used by leech
                         to suck its blood meal.

NOTE: The medicinal leech was extensively used in medieval medicine to remove excess blood from "feverish" or "sanguine" people. More recently, the medicinal leeches have been used as an integral part of microsurgery, where severed limbs and structures are resutured back into place. Leech saliva contains anticoagulants, anesthetics, and antibacterial proteins which are extremely useful in promoting blood flow and vein formation.

Here's a reply from Dr. Rudy Buntic, MD, a microsurgeon, regarding the use of leeches in microsurgery:

From: "Rudy Buntic" <rbuntic@microsurgeon.org>
To: "Dr. Stan Eisen" <seisen@cbu.edu>
Subject: Re: Do you use medicinal leeches in any of your microsurgery cases?
Date: Mon, 5 Nov 2001 19:49:26 -0800

Dear Dr. Eisen :

Thank you for your note. We do use leeches clinically, and fairly routinely. The most common indication is in finger replantation, when venous circulation is suggish or not present. Essentially, we examine the blood flow in a finger with two things in mind: arterial and venous circulation. If a replanted finger is nice and pink, then it has good arterial flow (inflow) and good venous flow (outflow). If a finger is white and therefore without good arterial inflow, then we don't usually find leeches helpful because there is no blood for the leeches to latch onto. In this case they cannot aid in the circulation of the finger. If the finger is bluish, dark and full of deoxygenated blood, then venous outflow is poor or non-existent and the finger will die. When venous circulation is compromised, a leech placed on the tip of the finger will cause a bleeding point that will allow blood to exit the finger, and therefore new oxygenated pink blood can nourish the finger. Often, within 7 to 10 days the finger will neovascularize with small capillaries, and then leeching will not be necessary because the capillaries will allow outflow of deoxygenated blood.

Another indication is microvascular transplants with a skin component. In these procedures, skin and subcutaneous tissue with muscle or bone is transplanted from one part of the body to another for reconstructive purposes. This is typically done with one artery and vein - similar to the situation in a replant. If venous circulation is problematic, we may use leeches.

I hope this is of some use to you. Let me know if you need any other information.

-Rudy Buntic, MD
Division of Microsurgery
45 Castro Street, Suite 140
San Francisco, CA 94114
415-565-6837
415-864-1654

And here's *fantastic* news, from the 9 July 2004 issue of Science

License to Leech

"The Food and Drug Administration (FDA) has approved the sale of leeches as a medical device. The French company that received FDA's thumbs up, Ricarimpex SAS, has been breeding leeches for 150 years. Although this is the first time FDA has explicitly granted permission for a company to sell them, there are plenty of homegrown leech vendors; several were grandfathered in under a 1976 law requiring the licensing of medical-device sellers.

Medical leeches, *Hirudo medicinalis*, are already used in plastic surgery to remove pooled blood from damaged areas, says Carl Krasniak of the Slocum-Dickson Medical Group of New Hartford, New York. The animals use a combination of chemicals in their saliva to prevent clotting and suck blood. FDA deemed them "devices" because their sucking action is considered more medically important than their anti-clotting saliva."

The following article appeared in the Commercial Appeal. The important paragraph states **"Sterile medicinal leeches were placed on the ear and scalp to suck away blood until veins can heal and begin pumping naturally."**

Rare surgery reattaches woman's scalp

LOS ANGELES (AP) — Doctors combined microsurgery and leeches to reattach the skin and hair of a woman who was scalped from her eyelids to the back of her neck by an industrial blender.

The accident at a packaging company threatened to leave 30-year-old Patsy Bogle disfigured. But doctors who performed the rare operation said Friday she would recover with few or no visible signs of her injury.

Bogle, an employee at Ross Technical Associates of Monrovia, 20 miles east of Los Angeles, said she was cleaning the blades of an industrial blender on Tuesday when the machine caught her ponytail. She felt her head smash into the machine. In an instant, her scalp was torn off from her eyelids to the back of her neck, including two-thirds of her right ear.

"It's not like if you lose a finger or thumb," Dr. John Gross said. "Those can be replaced, or you have other fingers to use. But a forehead, hair, eyebrows. . . Those would be very difficult — impossible — to reconstruct." She would have had to have extensive skin grafts and wear a wig the rest of her life.

Fortunately, her scalp came off in one piece. Paramedics removed it from the machine and packed it on ice to keep the tissue alive. Bogle was sent to specialists at the USC University Hospital.

During the five-hour operation, Gross and Dr. Bala Chandrasekhar reattached veins, muscle, cartilage and skin, including the ear. Sterile medicinal leeches were placed on the ear and scalp to suck away blood until veins can heal and begin pumping naturally.

Bogle will still need minor plastic surgery to fix her right eyelid. But swelling has started to go down, and doctors said she may be able to go home next week.

PHYLUM ACANTHOCEPHALA

CHARACTERISTICS OF MAJOR ANIMAL PHYLA

SYSTEM	TYPE/REMARKS
1. Circulatory	None. Exchange by Diffusion
2. Respiratory	None. Exchange by Diffusion
3. Excretory	Protonephridia With Flame Cells That Unite To Form A Common Tube That Opens Into Sperm Duct Or Uterus
4. Digestive	None. Absorption Is Through Tegument
5. Skeletal	None.
6. Nervous	Central Ganglion Within The Proboscis Receptacle
7. Type of Coelom	Pseudocoelomate
8. Muscular	Body Wall With Circular And Longitudinal Muscles
9. Reproductive	Egg layers, Dioecious

MAJOR GROUPS

CLASS EOACANTHOCEPHALA
CLASS PALAEACANTHOCEPHALA
CLASS ARCHIANTHOCEPHALA

Acanthocephalan sp.

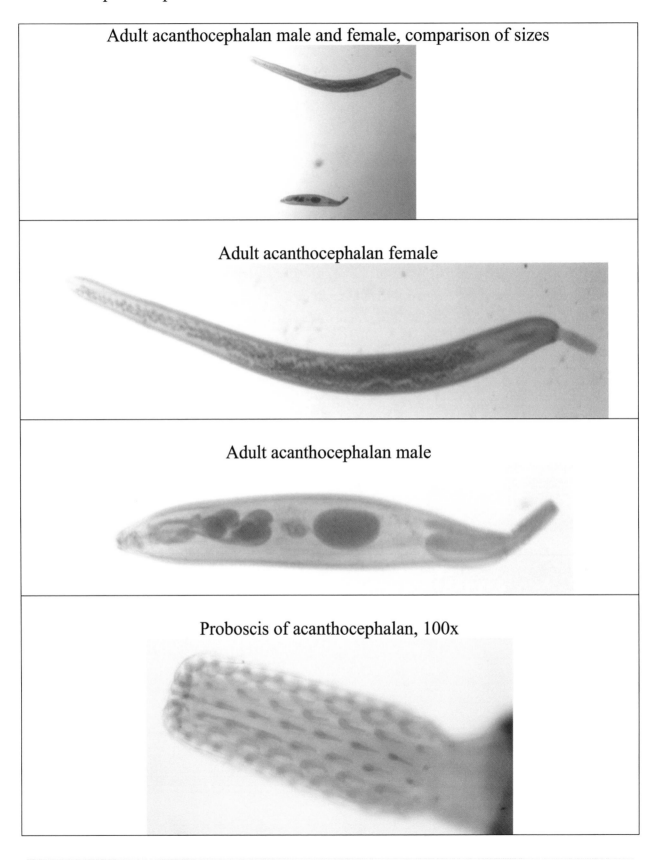

Adult acanthocephalan male and female, comparison of sizes

Adult acanthocephalan female

Adult acanthocephalan male

Proboscis of acanthocephalan, 100x

Macracanthorhynchus hirudinaceus

Images:
Ova:
http://www.k-state.edu/parasitology/625tutorials/Miscellaneous01.html

Phylogeny: Phylum Acanthocephala

Preferred definitive host: Hogs, wild boas, peccaries.

Reservoir hosts: To a limited degree, cats and dogs serve as hosts.
                            Humans are accidental hosts.

Vector/intermediate host: Larval beetles

Geographical location: Cosmopolitan. Human cases have been documented in
                            Bohemia and Russia on the basis of ova discovered
                            in the feces

Organs affected: Intestine

Symptoms and
Clinical Signs: Abdominal pain

Treatment: None indicated

Moniliformis moniliformis

Images:

Phylogeny: Phylum Acanthocephala

Preferred definitive host: Rats, mice, hamsters, dogs and cats

Reservoir hosts: Humans are incidental hosts

Vector/intermediate host: Beetles and cockroaches

Geographical location: Cosmopolitan. Human cases have been reported from Italy, the Sudan, and British Honduras

Organ affected: Small intestine

Symptoms and clinical signs: Abdominal pain, diarrhea, and exhaustion.

Treatment: None indicated.

Polymorphus paradoxus

Images:
Egg, acanthella, cystacanth, adult
http://www.bioscie.ohio-state.edu/~parasite/polymorphus.html

Phylogeny: Phylum Acanthocephala

Preferred definitive host: ducks

Reservoir hosts: None

Vector/intermediate host: *Gammarus* spp. amphipods

Geographical location: Cosmopolitan.

Organ affected: Small intestine

Symptoms and clinical signs: Generally none detectable. It may be fatal among ducklings.

Treatment: None indicated.

Note: A considerable amount of research has been conducted on the behavioral alterations on the intermediate hosts. Infected intermediate hosts are considerably more likely to be found by their predator, which is also the definitive host of the parasite.

        Comparative vulnerability of uninfected *Gammarus lacustris* and ones infected with *Polymorphus paradoxus*. (Adapted from Table 1 of Bethel and Holmes [1977], Increased vulnerability of amphipods to predation owing to altered behavior induced by larval acanthocephalans. Canadian Journal of Zoology 55(1):110-115.

Test No.	Duration (min)	No. predators	Gammarids eaten/No. available		P*
			Uninfected	Infected with *Polymorphus paradoxus*	
1	7	2	6/25	16/25	<0.01
2	5	2	13/50	35/50	<0.001
3	5	2	12/50	42/50	<0.001
4	5	1	8/50	18/50	<0.05
5	10	1	0/75	48/75	<0.001
6	15	1	24/75	63/75	<0.001
TOTAL			63/325	222/325	<0.001

*P = probability, by chi-square contingency tests

Phylum Mollusca
CHARACTERISTICS OF MAJOR ANIMAL PHYLA

SYSTEM	TYPE/REMARKS
1. Circulatory	Cephalopods Have A Closed System With A Chambered Heart. Hemocyanin and Hemoglobin Are Blood Pigments. Others have An Open System.
2. Respiratory	Aquatic Forms have Gills, Terrestrial Forms Have A Mantle Modified To Form A Lung
3. Excretory	Metanephridial Form. Principal Nitrogenous Waste Among Aquatic Forms Is Ammonia And Uric Acid Among Terrestrial Forms
4. Digestive	Complete
5. Skeletal	Shell May Be External, Internal Or Absent
6. Nervous	Ganglia Are Consolidated Around Esophagus -- Cerebral Ganglia
7. Type of Coelom	True, Reduced In Size
8. Muscular	Many Have Muscular Foot Or 'Pad' For Locomotion
9. Reproductive	Egg layers -- Most Are Dioecious

MAJOR GROUPS

CLASS POLYPHACOPHORA: CHITONS
CLASS GASTROPODA: SNAILS
CLASS SCAPHOPODA: MOLLUSKS WITH TUSK-LIKE SHELLS
CLASSBIVALVIA: OYSTERS AND BIVALVES
CLASS CEPHALOPODA: OCTOPUS, SQUID

Family Unionidae:

The larvae of unionid clams (called glochidia) are parasitic on the gills and skin of freshwater fish. Female clams attract fish and then "spray" larvae out of excurrent siphon. Valves clamp down on tissue, and a cyst of host-derived tissue forms around the larva. After the larvae reaches a certain size, it emerges from cyst and falls to bottom.

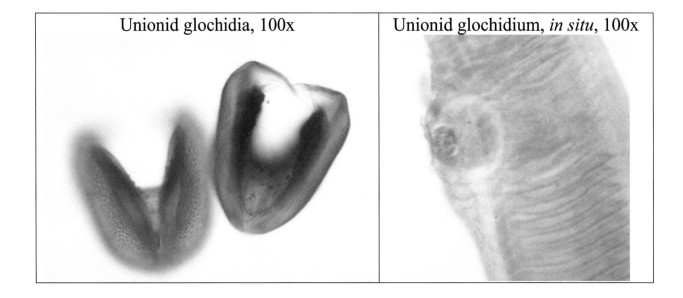

| Unionid glochidia, 100x | Unionid glochidium, *in situ*, 100x |

Phylum Arthropoda

CHARACTERISTICS OF MAJOR ANIMAL PHYLA
PHYLUM ARTHROPODA

SYSTEM	TYPE/REMARKS
1. CIRCULATORY	OPEN HEMOLYMPH FOUND IN MOST GROUPS HEMOGLOBIN FOUND IN LARVAE ADAPTED TO OXYGEN-POOR WATER
2. RESPIRATORY	SPIRACLES CHARACTERISTIC OF TERRESTRIAL FORMS, GILLS CHARACTERISTIC OF AQUATIC FORMS
3. EXCRETORY	MALPIGHIAN TUBULES AMONG TERRESTRIAL FORMS
4. DIGESTIVE	COMPLETE
5. SKELETAL	EXOSKELETON COMPOSED OF CHITIN
6. NERVOUS	ANTERIOR GANGLIA
7. TYPE OF COELOM	REDUCED IN SIZE, CALLED HEMOCOEL
8. MUSCULAR	STRIATED MUSCLE, JOINTED APPENDAGES
9. REPRODUCTIVE	MOST ARE DIOECIOUS. ISOLATED, SPECIFIC GROUPS SHOW PARTHENOGENESIS, POLYEMBRYONY

MAJOR GROUPS

CLASS MEROSTOMATA:	HORSESHOE CRABS
CLASS ARACHNIDA:	SPIDERS, TICKS, MITES, SCORPIONS
CLASS CRUSTACEA:	LOBSTER, CRABS, CRAYFISH, SHRIMP
CLASS DIPLOPODA:	MILLIPEDES
CLASS CHILOPODA:	CENTIPEDES
CLASS INSECTA:	INSECTS

PHYLUM ARTHROPODA

I. CLASS INSECTA

    A. ANOPLURA: SUCKING LICE
    B. DIPTERA: FLIES, MOSQUITOS, MIDGES
    C. HOMOPTERA: CICADAS, APHIDS
    D. HEMIPTERA: TRUE BUGS
    E. LEPIDOPTERA: MOTH AND BUTTERFLIES
    F. COLEOPTERA: BEETLES
    G. ORTHOPTERA: GRASSHOPPERS, COCKROACHES
    H. HYMENOPTERA: ANTS, BEES, WASPS
    I. ODONATA: DRAGONFLIES, DAMSELFLIES
    J. ISOPTERA: TERMITES

II. CLASS CRUSTACEA

    A. SUBCLASS BRANCHIOPODA: FAIRY SHRIMP, OR
    B. SUBCLASS OSTRACODA: SEED SHRIMP
    C. SUBCLASS COPEPODA:
    D. SUBCLASS BRANCHIURA: FISH LICE
    E. SUBCLASS CIRRIPEDIA: BARNACLES
    F. SUBCLASS MALACOSTRACA: WOOD LICE, PILL BUGS, SAND
                HOPPERS, CRABS, LOBSTER, CRAYFISH

III. CLASS ARACHNIDA

    A. ORDER SCORPIONIDAE: SCORPIONS
    B. ORDER ARANEAE: SPIDERS
    C. ORDER PHALANGIDA: DADDY-LONG-LEGS
    D. ORDER ACARINA: MITES AND TICKS
    E. SUBCLASS XIPHOSURA: HORSESHOE CRABS

IV. CLASS MYRIAPODA INCLUDES

    A. SUBCLASS CHILOPODA: CENTIPEDES
    B. SUBCLASS DIPLOPODA: MILLIPEDE

IMPORTANCE OF ARTHROPODS IN PARASITOLOGY

I. As parasites
  A. *Lernaea cyprinacea* among freshwater fish.
  B. *Argulus* among freshwater fish.
  C. *Unicola* mites among freshwater clams.

II. As parasitoids: (bloodsucking insects which then fly off the host.
  A. Mosquitos
  B. Horseflies
  C. Assasin bugs.

III. As Intermediate hosts.
  A. 1st. : *Dipylidium caninum.*
  B. 1st. : *Diphyllobothrium latum*
  C. 2nd. : *Paragonimus westermani*

IV. As vectors for disease.
  A. Mechanical transmission
    1. Cockroaches carry bacteria, feces on legs, mechanically carried to food.
  B. Biological transmission

V. As hyperparasites

Types of Biological Transmission

1. <u>Propagative biological transmission,</u> in which the disease-causing organism reproduces in the arthropod, but undergoes no further developmental changes.
    a. Plague bacillus in flea
    b. yellow fever virus

2. <u>Cyclopropagative biological transmission,</u> in which the disease-producing organism undergoes cylical changes and reproduces in the arthropod.
    a. *Plasmodium* in mosquitos
    b. *Trypanosoma* in tsetse flies

3. <u>Cyclodevelopmental biological transmission,</u> in which the disease-producing organism must undergo development in the arthropod but does not multiply there.
    a. Microfilarie in mosquitos

4. <u>Transovarian transmission,</u> in which disease-causing organisms are transmitted from the infected parent arthropod to their offspring.

CLASS INSECTA

- Found in terrestrial and aquatic habitats
- Three patterns of development:
  - Ametabolous – direct development
  - Hemimetabous – nymphal stages
  - Holometabolous – larval stages, followed by a pupal stage where there is complete rearrangement of tissues from imaginal disks

- Hormonal control of animal development
  http://www.ncbi.nlm.nih.gov/bookshelf/br.fcgi?book=dbio&part=A4302
- Exoskeleton provides framework for muscles, water conservation, armor, wings for flight.
- Dioecous
- Some species are social, with elaborate colonies
  - Hymenoptera (bees and ants)
  - Isoptera (Termites)

Order Diptera

FAMILY CULICIDAE (MOSQUITOS), INCLUDING GENERA *Anopheles, Culex, Mansonia, Aedes*

Images:

Adult *Culex* emerging from pupal case:
http://www.k-state.edu/parasitology/625tutorials/Arthropods09.html

Phylogeny:	Order Diptera
Metamorphosis:	Complete. Larvae are aquatic.
Geographical location:	Cosmopolitan
Organs affected:	Skin
Symptoms and clinical signs:	Bite is followed by erythema, swelling and itching.
Diseases transmitted:	Yellow fever, dengue, viral encephalitis, filariasis, malaria.
Treatment/control:	Residual spraying, drainage of marsh or swamp areas, covering of cisterns with diesel oil or covers. Biological control of larvae is accomplished with predaceous fish such as *Gambusia* (mosquitofish).

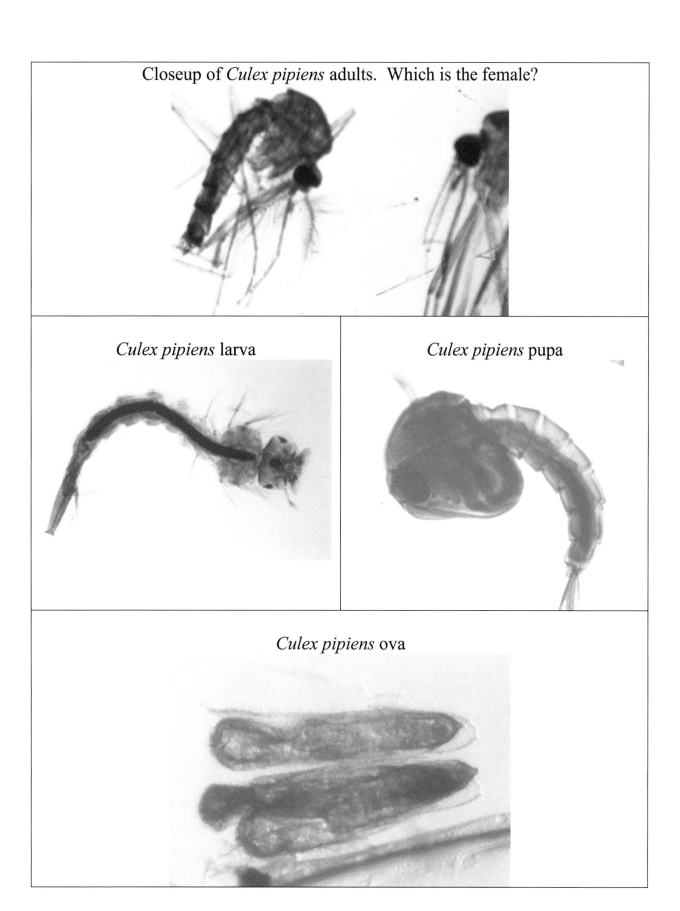

Closeup of *Culex pipiens* adults. Which is the female?

Culex pipiens larva

Culex pipiens pupa

Culex pipiens ova

Simulium spp. (BLACKFLIES)

Phylogeny: Order Diptera

Metamorphosis: Complete

Geographical location: Cosmopolitan

Organs affected: Skin

Symptoms and clinical signs: Bites, which are painless at first, bleed profusely. Swelling, pruritis, and pain develop later.

Diseases transmitted: Onchocerciasis

Treatment/control: Residual insecticide

Glossina spp. (TSETSE FLIES)

Images:
Wing structure:
http://www.k-state.edu/parasitology/625tutorials/Wings02.html

Phylogeny:	Order Diptera
Metamorphosis:	Complete
Geographical location:	Africa South of Sahara
Organs affected:	Skin
Symptoms and clinical signs:	Dermal irritation
Diseases transmitted:	Nagana and African sleeping sickness.
Treatment/control:	Residual insecticides, destruction of brush used for breeding.

Glossina sp. adult

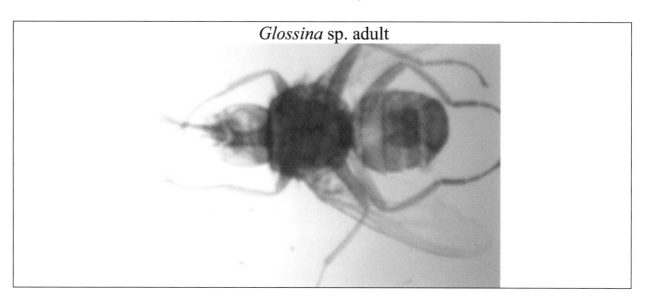

Chrysops spp. (DEERFLY)

Images:

Adult horsefly (*Tabanus* sp.)

Phylogeny:	Order Diptera
Metamorphosis:	Complete
Geographical location:	Cosmopolitan, but abundant in Americas
Organs affected:	Skin
Symptoms and clinical signs:	Dermal irritation and hemorrhage of blood through wound.
Diseases transmitted:	*Loa loa*
Treatment/control:	Soothing lotion to treat symptoms, residual insecticide, larvicides.

Phlebotomus spp. (SANDFLIES)

Phylogeny: Order Diptera

Metamorphosis: Complete

Geographical location: Tropical and Subtropical Countries of the Old World

Organs affected: Skin

Symptoms and
clinical signs: Rose-colored papules and stinging
                          pain at site of bites.

Diseases transmitted: Leishmania donovani, Leishmania braziliensis, sandfly
                          fever (viral), and Bartonellosis (bacteria).

Treatment/control: Residual insecticides.

Lutzomyia spp. (NEW WORLD SANDFLIES)

Phylogeny: Order Diptera

Metamorphosis: Complete

Geographical location: Tropical and subtropical areas of
                          the New World.

Organs affected: Skin

Symptoms and clinical signs: Rose-colored papules and stinging pain at site of
                               bites.

Diseases transmitted: *Leishmania donovani, Leishmania braziliensis,*
                               sandfly fever (viral), and Bartonellosis
                               (bacterial).

Treatment/control: Residual insecticides.

Order Hemiptera:

Triatoma infestans

Images:
Adults:

Phylogeny:	Order Hemiptera
Metamorphosis:	Incomplete
Geographical location:	South America
Organs affected:	Skin
Symptoms and clinical signs:	Bites are often symptomless
Diseases transmitted:	*Trypanosoma cruzi*
Treatment/control:	Insecticides and replacement of thatched roofs with sheet metal.

Rhodnius prolixus

Images:

Phylogeny:	Order Hemiptera
Metamorphosis:	Incomplete
Geographical location:	South America
Organs affected:	Skin
Symptoms and clinical signs:	Bites are frequently symptomless. occasionally, victims will have pruritic skin reactions
Diseases transmitted:	*Trypanosoma cruzi*
Treatment/control:	Lindane, dieldrin. Spraying of juvenile hormone appears promising as a means of control.

Cimex spp. INCLUDING *C. lectularis* and *C. hemipterus* (bedbugs)

Adult:
http://www.k-state.edu/parasitology/625tutorials/Arthropods02.html

Phylogeny:	Order Hemiptera
Metamorphosis:	Incomplete
Geographical location:	*C. lectularis* is cosmopolitan, whereas *C. hemipterus* is found in West Africa
Organs affected:	Skin
Symptoms and clinical signs:	Bites are frequently symptomless, but they may disturb sleep, reduce hemoglobin, or induce inflammation.
Diseases transmitted:	None
Treatment/control:	Residual insecticides. itching is relieved with calamine lotion.

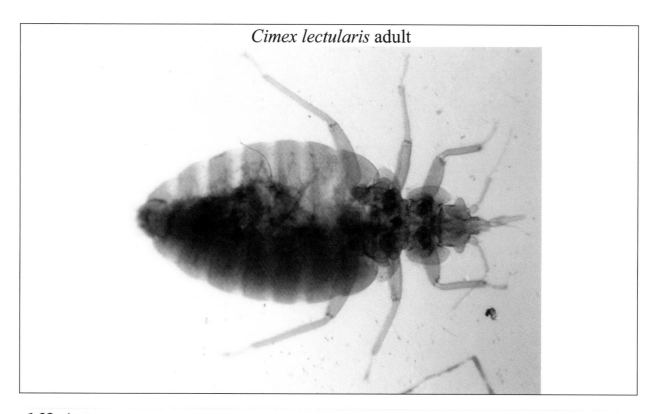

Cimex lectularis adult

Order Siphonaptera - Fleas

Pulex irritans

Images:

Phylogeny: Order Siphonoptera

Metamorphosis: Complete

Geographical location: Europe and Western United States

Organs affected: Skin

Symptoms and clinical signs: Itching dermatitis

Diseases transmitted: None

Treatment/control: Environmental control with insecticides

Pulex irritans adult male

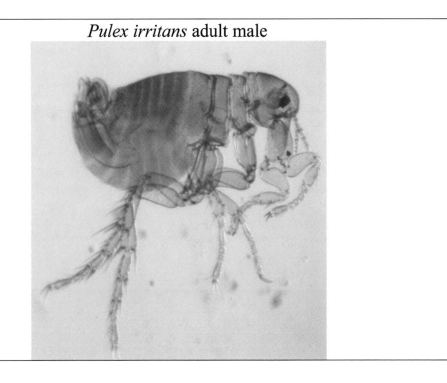

Ctenocephalides spp. including *C. canis* and *C. felis*

Adult:
http://www.k-state.edu/parasitology/625tutorials/Arthropods02.html
Line drawing of adult:
http://www.k-state.edu/parasitology/625tutorials/Arthropods07.html

Phylogeny: Order Siphonoptera

Metamorphosis: Complete

Geographical location: Cosmopolitan

Organs affected: Skin

Symptoms and clinical signs: Itching dermatitis

Diseases transmitted: *Dipylidium caninum, Dirofilaria immitis,*
           Dipetalonema reconditum

Treatment/control: Residual insecticides and maintenance of a clean
           environment.

Ctenocephalides sp. adult female

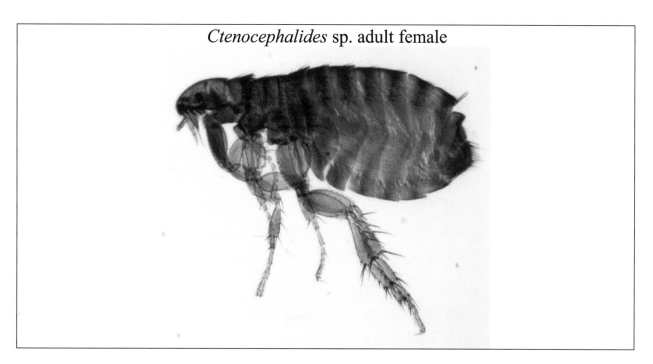

Tunga penetrans

Images:
Adult, infected foot

Phylogeny:	Order Siphonoptera
Metamorphosis:	Complete
Geographical location:	Tropical America, parts of Africa, Near East, India
Organs affected:	Skin
Symptoms and clinical signs:	Fertilized female burrows into the skin of mammals or humans. Lesions can become a festering, painful sore. Secondary bacterial infections are common.
Diseases transmitted:	None
treatment/control:	Tropical DDT treatment. Burrowing females are surgically removed.

Xenopsylla cheopis (RAT FLEA)

Image:
Adults:
http://www.k-state.edu/parasitology/625tutorials/Arthropods03.html

Phylogeny:	Order Siphonoptera
Metamorphosis:	Complete
Geographical location:	Cosmopolitan on *Rattus* spp. rats except in cold climates.
Organs affected:	Skin
Symptoms and clinical signs:	Dermal irritation.
Diseases transmitted:	Bubonic plague. Symptoms of plague include swollen lymph nodes, hemorrhage, mental dullness. Within a relatively short time, patient shows anxiety, delirium, coma and death.
Treatment/control:	Antirat campaigns must be preceded by a spraying program to eradicate fleas. Antibiotics are effective against plague.

Order Anoplura – Lice

Pediculus humanus (body and head lice)

Adults:
http://www.k-state.edu/parasitology/625tutorials/Arthropods05.html

Phylogeny:	Order Anoplura (sucking lice)
Metamorphosis:	Incomplete
Geographical location:	Cosmopolitan
Organs affected;	Skin
Symptoms and clinical signs:	Saliva induces roseate elevated papules. Severe infestation lead to scarring, induration, ulceration.
Diseases transmitted:	Epidemic typhus, trench fever, relapsing fever
Treatment/control:	Head lice: Shampoo with pyrethrins (0.2%), piperonyl butoxide and copper oleate. If that doesn't work, use olive oil or mayonnaise, leave on head overnight. Brush hair thoroughly. Body lice: Shampoo containing 0.2% or 0.3% allethrin synergized with piperonyl butoxide.

Pediculus humanus adult male and female

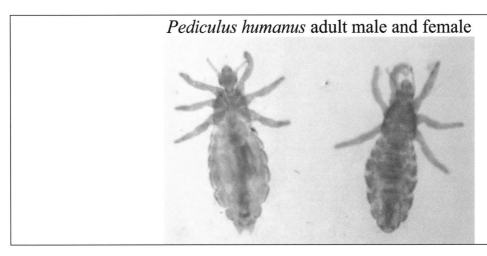

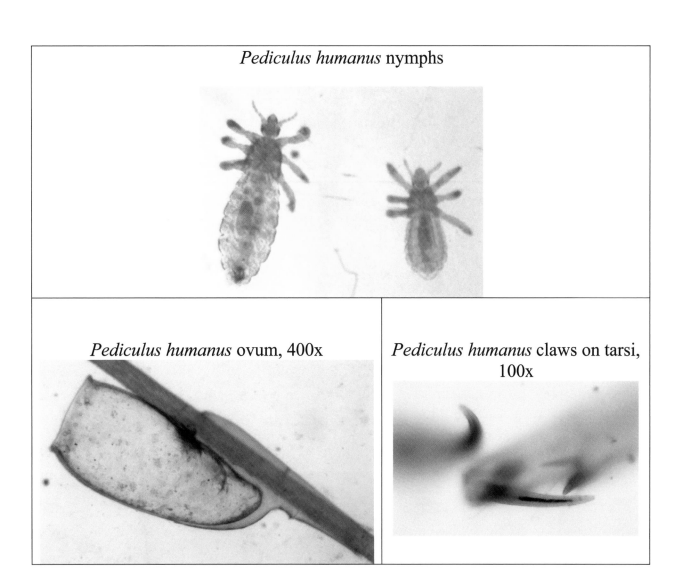

Pediculus humanus nymphs

Pediculus humanus ovum, 400x

Pediculus humanus claws on tarsi, 100x

From: http://www.historyhouse.com/in_history/lousy/

During the Russian revolution, there was an outbreak of typhus (transmitted by lice) so severe that Lenin remarked, **"Either socialism will defeat the louse, or the louse will defeat socialism."**

To get an idea of how powerful a force disease is, and to remind the historian that it should not be overlooked, allow us to quote Hans Zinssner's account of a famous plague of ancient times -- the Plague of Justinian. It started in the year 540, perhaps prompted by a series of earthquakes and floods which created refugee conditions across much of Eastern Christendom.

From Hans Zinsser's Rats, Lice and History (copyright

1941)

Four months the plague remained in Byzantium. At first, few died -- then there were 5000, later **10,000 deaths a day**. [Quoting from Procopious, a contemporary historian. Such numbers are almost surely exaggerations, as any number above a few thousand tended to mean 'many' in those times - HH] 'Finally, when there was a scarcity of gravediggers, the roofs were taken off the towers of the forts, the **interiors filled with the corpses**, and the roofs replaced.' Corpses were placed on ships, and these abandoned to the sea. 'And after the plague had ceased, there was so much depravity and general licentiousness, that it seemed as though **the disease had left only the most wicked**.'

Also From Hans Zinsser's Rats, Lice and History (copyright 1941)

...among the Aztecs before the advent of Cortez, is the tale cited from Torquemada. 'During the abode of Montezuma among the Spaniards, in the palace of his father, Alonzo de Ojeda one day espied... a number of small bags, tied up. He imagined at first that they were filled with gold dust, but on opening one of them what was his astonishment to find it **quite full of Lice**!' Cortez... then asked... for an explanation. He was told that the Mexicans had such a **sense of duty** to pay tribute to their ruler that the poorest, if they possessed nothing else to offer, daily **cleaned their bodies and saved the lice**. And when they had enough to fill a bag, they laid it at the feet of their king.

MacArthur's story of Thomas a Becket's funeral illustrates [this]: -- The archbishop was murdered in Canterbury Cathedral on the evening of the twenty-ninth of December. The body **lay in the Cathedral all night**, and was prepared for burial on the following day... He had on a large brown mantle; under it, a white surplice; below that, a lamb's-wool coat; then another woolen coat; and a third woolen coat below this; under this, there was the black, cowled robe of the Benedictine Order; under this, a shirt; and next to the body a curious hair-cloth, covered with linen. As the body grew cold, the vermin that were living in this multiple covering started to crawl out, and, as MacArthur quotes the chronicler: 'The vermin boiled over like water in a simmering cauldron, and the onlookers burst into **alternate weeping and laughter**.'

Robert Burns' Ode to a Louse, appearing at
http://forums.eslcafe.com/student/viewtopic.php?p=738

Robert Burns (1759-1796)

TO A LOUSE, ON SEEING ONE ON A LADY'S BONNET AT CHURCH

Ha! whare ye gaun, ye crowlan ferlie!
Your impudence protects you sairly;
I canna say but ye strunt rarely,
Owre gauze and lace;
Tho', faith! I fear ye dine but sparely
On sic a place.

Ye ugly, creepan, blastit wonner,
Detested, shunn'd by saunt an' sinner,
How daur ye set your fit upon her,
Sae fine a Lady!
Gae somewhere else and seek your dinner
On some poor body.

Swith! in some beggar's haffet squattle;
There ye may creep, and sprawl, and sprattle,
Wi' ither kindred, jumping cattle,
In shoals and nations;
Whare horn nor bane ne'er daur unsettle
Your thick plantations.

Now haud you there, ye're out o' sight,
Below the fatt'rels, snug and tight,
Na, faith ye yet! ye'll no be right,
Till ye've got on it,
The verra tapmost, towrin height
O' Miss's bonnet.

My sooth! right bauld ye set your nose out,
As plump an' grey as onie grozet:
O for some rank, mercurial rozet,
Or fell, red smeddum,
I'd gie you sic a hearty dose o't,
Wad dress your droddum!

I wad na been surpriz'd to spy
You on an auld wife's flainen toy;
Or aiblins some bit duddie boy,

On's wylecoat;
But Miss's fine Lunardi, fye!
How daur ye do't?

O Jenny, dinna toss your head,
An' set your beauties a' abread!
Ye little ken what cursed speed
The blastie's makin!
Thae winks and finger-ends, I dread,
Are notice takin!

O wad some Pow'r the giftie gie us
To see oursels as others see us!
It wad frae monie a blunder free us,
An' foolish notion:
What airs in dress an' gait wad lea'e us,
And ev'n Devotion!

Here is THE quote from Hans Zinsser's Rats, Lice and History (copyright 1941):

"Weizl (*an Austrian anthropologist*) informs us that, when sojourning for a short time among the natives of Northern Siberia, young women who visited his hut sportively threw lice at him. On inquiry concerning this disconcerting procedure, he was embarrassed by learning that this was the customary manner of indicating love, and a notice of serious intentions. A sort of 'My louse is thy louse' ceremony."

Phthirus pubis (crab lice)

Images:

Adult:
http://www.k-state.edu/parasitology/625tutorials/Arthropods04.html

Phylogeny:	Order Anoplura (sucking lice)
Metamorphosis:	Incomplete
Geographical location:	Cosmopolitan
Organs affected:	Skin, particularly of the pubic region.
Symptoms and clinical signs:	Saliva induces roseate elevated papules. Severe infestations lead to scarring, ulceration.
Diseases transmitted:	None
Treatment/control:	On pubic area, treat as for head lice. Nits and lice may be removed from eyelashes with forceps. Ophthalmic ointments of Eserine or of yellow oxides of mercury are both effective.

Phthirus pubis adult

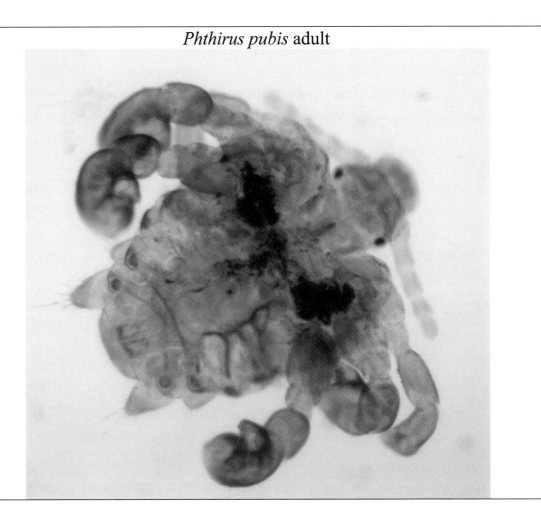

Phthirus pubis claws, 100x

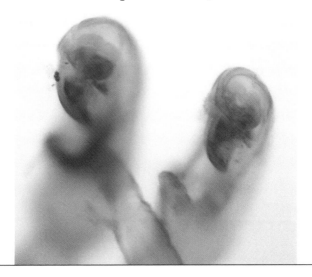

Ticks and other Acarines:

Tick Testing Centers

North American Laboratory
New Britain, CT
203-826-1140
800-866-NALG
203-223-6279-fax

New Jersey Laboratories
New Brunswick, NJ
732-249-0148

IgeneX
Palo Alto, California
http://www.igenex.com/
415-424-1191
800-832-3200

Tick Research Laboratory
Kingston, Rhode Island
401-874-2650

Dermacentor andersoni (wood tick)

Images:
Adult *Dermacentor variabilis:*
http://www.k-state.edu/parasitology/625tutorials/Arthropods12.html

Phylogeny: Class Arachnida

Metamorphoosis: 'Incomplete'.   Larvae and nymphs resemble adults.

Geographical location: North America

Organs affected: Skin

Symptoms and clinical signs: Inflammation, edema, hemorrhage, secondary
                                            bacterial infection, tick paralysis.

Diseases trasmitted: American spotted fever (rickettsia), Q fever
                                            (rickettsia), Colorado tick fever (virus),
                                            Viral encephalitis, Tularemia (bacteria).

Treatment/control: Topical insecticide and use of repellants on
                                            clothing.

Dermacentor andersoni adult female

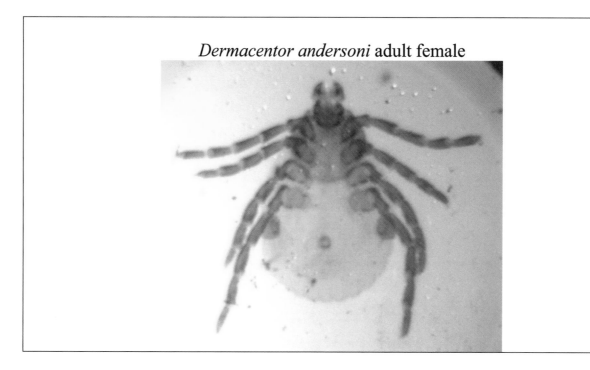

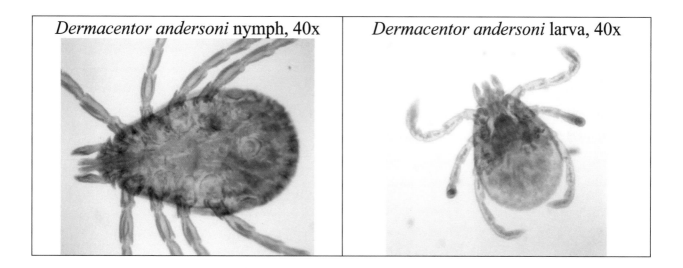

Dermacentor andersoni nymph, 40x

Dermacentor andersoni larva, 40x

Ixodes spp., *Boophilus* spp., *Amblyomma* spp. (HARD TICKS)

Images:

Ixodes scapularis (black-legged tick) adults:
http://www.k-state.edu/parasitology/625tutorials/Arthropods22.html

Phylogeny:	Class Arachnida
Metamorphosis:	'Incomplete". Larvae and nymphs resemble adults.
Geographical location:	Cosmopolitan
Organs affected:	Skin
Symptoms and clinical signs:	Inflammatory responses, including local hyperemia, edema, hemorrhage.
Diseases transmitted:	American spotted fever (rickettsia), Viral encephalitis, Tularemia (bacteria), Babesia (protozoa).
Treatment/control:	Topical insecticide and use of repellents.

Trombicula alfreddugesi (CHIGGERS)

Images:

Trombiculidae, from *Rana pipiens*	Chigger mite, presumably from human, 100x
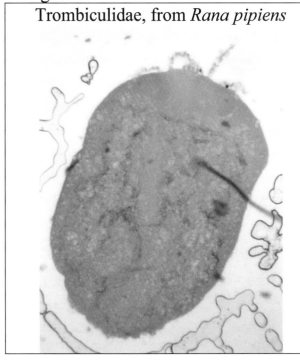	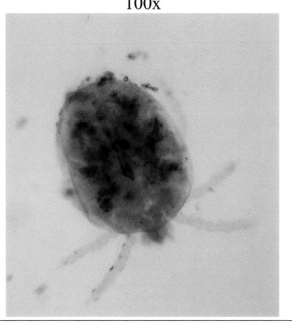

Phylogeny: Class Arachnida

Metamorphosis: 'Incomplete'. larvae and nymphs resemble adults.
                                larvae are parasitic.

Geographical location: North America and Europe

Organs affected: Skin

Symptoms and clinical signs: Bite causes swelling and intense itching. Infection
                                    may be debilitating due to loss of sleep.

Diseases transmitted.   Tsutsugamushi disease.

Treatment/control: Hot soap and water bath followed by the
                                    application of a 10% sulfur ointment
                                    containing 1% phenol relieve itching.
                                    Residual insectides and repellents are used.

Sarcoptes scabiei (SCABIES), *Notoedres cati* (FACIAL MANGE in cats)

Images:

Adults:
http://www.k-state.edu/parasitology/625tutorials/Arthropods01.html
More adults:
http://www.k-state.edu/parasitology/625tutorials/Arthropods15.html

Phylogeny:	Class Arachnida
Metamorphosis:	'Incomplete'. Larvae and nymphs resemble adults.
Geographical location:	Cosmopolitan
Organs affected:	Skin
Symptoms and clinical signs:	Lesions appear as reddish slightly elevated tracts in the skin. Intense itching causes scratching, leading to secondary infections.
Diseases transmitted:	None
Treatment/control:	Ointment containing 1% gamma benzene hexachloride. Pyrethrins are also used and are less toxic.

Demodex spp. including *D. folliculorum* and *D. canis*

Images:

Companion animal *Demodex* spp.:
http://www.k-state.edu/parasitology/625tutorials/Arthropods17.html
More adults:
http://www.k-state.edu/parasitology/625tutorials/Arthropods08.html

"Eyelash Creatures":
http://www.geocities.com/thesciencefiles/eyelash/creatures.html

Phylogeny:	Class Arachnida
Metamorphosis:	'Incomplete'. Larvae and nymphs resemble adults.
Geographical location:	Cosmopolitan
Organs affected:	Hair follicles and sebaceous glands
Symptoms and clinical signs:	Acne, blackheads, localized keratitis. *Demodex* induces mange among dogs.
Diseases transmitted:	None
Treatment/control:	Treatment is rarely required for human infections.

PHYLUM CHORDATA

Cuckoos and other brood parasites:

From **Nature's Camouflage,** by Edith Banks, Albany Books, 1979:

"An interesting sort of mimicry is found in cuckoos: not in the birds themselves, but in their eggs. As is well known, many cuckoos lay their eggs in the nests of other birds which then rear the baby cuckoo as if it were their own. Not all cuckoo species do this. Some build a nest and raise their own young, and these provide an interesting comparison with the parasitic cuckoos.

The non-parasitic species usually lay quite large, plain, white eggs. The parasitic species almost all lay smaller eggs, and these are coloured and patterned to mimic the eggs of the birds who unwittingly adopt them. Many different birds are used as foster-parents by cuckoos, but generally, in any one region, a cuckoo species parasitizes one particular bird, and the cuckoo's eggs specifically imitate the eggs of that bird.

If the nest belongs to a large, intelligent species, such as a Crow or a Magpie, the egg mimicry is very good. Smaller birds are not so intelligent, and can be duped just as easily by an egg which is only an approximate copy of their own. Most can detect that an egg is an alien one if the pattern and colour are very different, and will eject it from the next. But they do not seem perturbed by an egg which is larger than their own. Often the pattern and colouring are the same but the cuckoo's eggs are half as large again as those of the foster parent.

It may even be that birds prefer larger eggs. Experiements with Herring Gulls (which, incidentally, are not parasitized by cuckoos) , have shown that this is so for them. If a model of the Herring Gull's egg, correct in colour, shape and pattern but three times larger than normal, is offered to a female gull, she will brood it in preference to her own. The egg is acting as a 'super-stimulus'. The large, fake egg compels the gull to incubate it far more strongly than one of her own, much smaller eggs.

The same mechanism is at work when small birds which have hatched a cuckoo's egg continue to feed the nestling even though it grows to twice their size. A tiny Reed Warbler will persist in feeding even though it may have to perch on the back of the grotesquely large cuckoo chick in order to reach its beak.

So powerful is the sight of the cuckoo chick's open beak that another small bird passing by will respond to it, although it has never seen the chick before. The passing bird will suddenly stop and push into the greedy mouth a morsel of food which was destined for its own young ones.

A cuckoo which is fostered by the more intelligent Crows and Magpies needs to be more restrained. IT has to share the nest with the foster bird's own young (those that parasitize small birds just push the other eggs out of the nest as soon as they can). But since these larger birds are less gullible the cuckoo chick needs to mimic one of their own nestlings in order to be fed. Such cuckoo chicks have evolved colouring on the tops of their heads, on their backs and inside their beaks which mimics that of the other young birds. Their undersides, however, look quite different, but this does not matter since the parent bird never sees them from below. "

I. Cuckoos do not produce their own nests, but lay eggs in nests produced by other species.
II. Eggs laid by cuckoos resemble those of "foster" parents in size and coloration pattern, but are generally larger. Larger size allows for earlier hatching. Sometimes, foster parents can tell, in their own "birdy" way, that something is not right with their nest, so they will toss out one egg. Because of hyperstimulation by the cuckoo's egg, they are more likely to toss out one of their own.
III. Cuckoo nestlings show similar "begging" behavior for eliciting food from foster parents. Cuckoos start getting fed even before the foster parents' young are hatched.
IV. Cuckoo nestlings push foster nestlings out of the nest, so the cuckoo nestling is often the only remaining one in the nest. (This is not pleasant to watch.)
V. Cuckoos are often found along borders of different communities, so native bird populations, already reduced because of fragmentation of their habitats, are getting clobbered.

References:
        Payne, Robert B. and Laura L. Payne. 1997. Brood parasitism by cowbirds: risks and effects on reproductive success and survival in indigo buntings. Behavioral Ecology 9(1):64-73.

        Payne, Robert B. 1998. Brood parasitism in birds: Strangers in the nest. Bioscience May 1998.

Petromyzon marinus (sea lamprey)

Phylogeny:	Class Agnatha
Metamorphosis:	Complete
Geographical location:	Over the past 100 years, populations have succeeded in bypassing the marine stage, so that they can live their entire lives in freshwater. Adults live in deep waters of the Great Lakes, and migrate to tributaries & rivers to spawn. Ammocoete larvae live upstream for several years, and then migrate to open water, where they reach sexual maturity.
Organs affected:	Skin. Rows of numerous teeth cause severe damage to skin, thereby promoting secondary bacterial & fungal infections which can be fatal.
Treatment/control:	Larvicides are applied in small rivers and streams

Vandellia cirrhosa AKA *Candiru*

Image:
http://www.k-state.edu/parasitology/625tutorials/Candiru.html

(Description is from
http://www.sciencenet.org.uk/database/Biology/0003/b00768d.html

I have heard about a tropical fish which can detect traces of mammalian urine in the water and will enter the urinary tract of the mammal. Is this true? What is it's taxonomy etc.?

The fish you are referring to is the Vandellia cirrhosa, common name candiru, a member of the family Trichomycteridae the pencil or parasitic catfishes. Vandellia is about an inch (2.5 cm) in length and when it has not fed is slender and almost transparent (except for the eyes). It lives in the rivers of tropical South America.This small catfish is a vampire - it feeds on the blood of other fish.

It has been described as entering the gill chambers of larger fish to suck blood from their gills. Once in the gill chamber it anchors itself there, so as not to be flushed out as the fish pumps water over its gills, with spines on its gill covers. As it feeds the body becomes engorged and distended with blood. Once it has fed the candiru swims out of the gill chamber and burrows into the river-bed to digest its blood meal.

You are correct in believing that the candiru poses a hazard to humans (and other mammals that might urinate in the water). It seems attracted to the flow of urine (possibly as it resembles the stream of water from the gills of a large fish). The candiru may swim up the stream of urine and enter the urethra of a bather urinating into the river.

This, of course, not part of the fish's normal feeding behaviour - the fish has made a fatal mistake. Once up the urethra the fish can not turn nor can it move backwards because of the rear-pointing spines on its gill covers. It is locked in. The fish invariably dies and the dead fish and associated swelling of the lining of the urethra cause the urethra to become blocked. Surgery is required to remove the obstruction.

Some recent expeditions to the Amazon region have had their personnel wear cricket box type shields to protect against candiru whilst swimming in or wading through streams. Authorities in the USA are concerned that candiru might possibly escape from aquaria and populate the rivers of the southern US. For this reason the importation into parts of the USA of any member of the family Trichomycteridae is strictly prohibited.

http://www.internext.com.br/urologia/Casosclinicos.htm

Plant Kingdom

From: http://www.gardenbuildingsdirect.co.uk/Article/parasitic-plants

Parasitic plants use the water and nutrients of their host plants to sustain themselves. They generally don't kill their hosts; however, they don't benefit the hosts either. A parasitic plant has a special type of root, called the haustorium, which it uses to pierce through the host's tissues and connect to the xylem or phloem.

More than four thousand species of parasitic plants have been identified to date, and they belong to 19 different flowering plant families. There are 6 major types of parasitic plants, and they are the holoparasites, hemiparasites, obligate parasites, facultative parasites, stem parasites, and the root parasites. Holoparasites are plants that do not have chlorophyll, and they cannot survive without parasitism, whereas hemiparasites can photosynthesize partially, and hence, they can live as parasites or on their own. Obligate parasites, on the other hand, require a host to complete a life cycle, but in the case of facultative parasites, a host isn't needed for completing the life cycle. A stem parasite will attach itself to the stem of a host, while the root parasite will attach itself to the root.

Some of the most common parasitic plants are the several species of mistletoes that belong to the loranthaceae and viscaceae families. In fact, they make up about 75% of all parasitic plant species. Dodders cuscutaceae and broomrapes orobanchaceae are two other families of common parasitic flowering plants, and they are often found around summerhouses.

There are some general characteristics of parasitic plants. First of all, parasitic plants use the haustorium to transport nutrients and water. Secondly, they have the tendency to completely abandon their own photosynthesis sometimes. Also, parasitic plants called epiparasites can be exposed at the surface of a host's skin, while endoparasites can be concealed within the host's organ. Finally, parasitic plants use their vascular systems to connect to their hosts.

A parasite establishes itself through germination, and it quickly develops a modified lateral root, which is the haustorium. A disc shaped structure called the hapteron is then formed, which helps the root tip to penetrate the host and connect its own phloem or xylem to the host's phloem or xylem.

Parasitic plants are not always despised by people. For instance, nuytsia and broomrape are parasitic plants which are cultivated in greenhouses for their beautiful flowers. There are certain parasitic plants that are used by researchers to study the evolution of photosynthesis. Moreover, about a dozen of them can be consumed as food or medicine. However, some parasitic plants are poisonous, and they may cause a variety of health problems when consumed, including intestinal problems and a weak pulse. It has been shown that consumption of the American mistletoe may result in death. There are also parasitic plants that can cause skin irritations when they are touched. Parasitic plants are generally considered detrimental to the economy, since they damage economically important plants as well as crops. The foliage of the parasites dodder and mistletoe covers the host to such an extent that almost all of the host's foliage is not visible.

· The Parasitic Plant Connection : Detailed information on parasitic plants

· Britannica : Definition of haustorium

· UCLA Botany – Parasitic Plants : Provides information about parasitic plants

· Forms of Parasitism : Identifying the different types of parasitic plants

· Wikipedia - Mistletoe : Information on mistletoe

· Colorado State University - Dodder : Comprehensive information about dodders

· Broomrape : Information on broomrapes

· Penn State University – Parasitic Plants and Their Hosts : Research on how parasitic plants find their hosts

· ASGAP.org - Nuytsia : Information about the Nuytsia plant

· Wikipedia - Rafflesia : Information on Rafflesia, a parasitic plant with the world's largest flower

· Access Science – Parasitic Plants : Research on parasitic plants

· National Center for Biotechnology Information – Photosynthesis of Parasitic Plants : Information about the photosynthesis process in parasitic plants

· <u>Parasitic Plant Food</u> : List of parasitic plants that can be used as food

· <u>ITM Online.org – Cynomorium</u> : Information on Cynomorium, a parasitic plant used in traditional medicine

· <u>Botany.org – Parasitic Plants</u> : Pictures of parasitic plants

Although the botanical features of parasitic plants such as their evolution, physiology, and anatomy have been studied in detail, not much in-depth information on the impact of parasitic plants on their hosts has been gathered, specifically from the perspective of damage caused. Nonetheless, experts agree that parasitic plants rarely kill their hosts, but they also believe that parasites and their hosts do not live happily together.

From the 30 March 2007 issue of Science (Volume 315)

Rafflesia tuan-mudae

A bloom of *Rafflesia tuan-mudae* from western Borneo. Rafflesiaceaea species produce the world's largest flowers, with blooms up to 1 meter in diameter. Molecular evidence shows that these enigmatic parasites are members of the spurge family, Euphorbiaceae (which, by the way, includes *Poinsetta* sp.), hence the enormous flowers of Rafflesiaceae most likely arose from tiny-flowered ancestors.

Monotropa uniflora -- Indian pipes

From: <u>http://botit.botany.wisc.edu/toms_fungi/oct2002.html</u>

This month's fungus is not a fungus at all, but is often brought in to forays and by students thinking it must be a fungus because it's white and doesn't have any chlorophyll. But it's really a flowering plant-- in the blueberry family! This is one of about 3000 species of non-photosynthetic (i.e. heterotrophic) flowering plants. How does this plant survive?? I'll tell you later of the interesting way that this non-photosynthetic plant gets its food.

Monotropa uniflora can actually grow in dark (and spooooooooooky) environments because it is not dependent on light for photosynthesis. I tend to find this plant in rich habitats-- dense moist forests with much surface leaf litter, often in a situation that is too shaded for autotrophic (photosynthetic) growth. Finding the ghost plant is an indication to me that I am in a very rich woods, and I should be on the lookout for lots of interesting fungi. *Monotropa uniflora* is the most common species in Wisconsin and the rest of North America east of the Great Plains. It is also known from Japan, and probably occurs in other places as well. There are relatives of this plant that occur throughout the world.

There are at least 3000 species of non-photosynthetic members of the plant kingdom. All of these are vascular flowering plants (angiosperms), except for one weird non-photosynthetic liverwort that I know of (*Cryptothallus mirabilis*). Many of these angiosperms are members of the Ericaceae, a family that also includes blueberries, cranberries, heath, *Rhododendron*, azaleas, *Arctostaphylos*, and *Arbutus*. There are many other species of *Monotropa*, as well as other genera of mycoparasitic plants including *Pterospora, Hermitomes, Sarcodes, Pityopus* and others. All of these non-photosynthetic members of the Ericaceae belong to the subfamily Monotropoideae. There are a number of other plants in other families that I will discuss later.

Since all of these plants are heterotrophic, they must get their food from an outside source. Almost all are parasitic on other organisms. Many (like mistletoe and dwarf mistletoe) are directly parasitic on other plants. However, most of these heterotrophic plants, and certainly all of the monotropes are parasitic on fungi! These fungi are mycorrhizal with photosynthetic trees, and thus the energy ultimately comes from photosynthesis of the tree, passing through the mycorrhizal fungus on the way to the *Monotropa*.

- The tree, already providing energy to the fungus, is probably physiologically 'unaware' of the additional loss of carbon and it is likely that it is the fungus that controls the passage of carbon to Monotropa. The one-way flow of carbohydrates can be traced by supplying the photosynthetic tree with radioactive carbon dioxide, i.e. $^{14}CO_2$.

Carbon dioxide is fixed into sucrose, which is transported to the roots of the tree. The mycorrhizal fungus takes the sucrose and transforms it into trehalose or sugar alcohols, which are transported to the rest of the fungal mycelium. (In return the fungus aids the tree in absorption of water and essential minerals, especially phosphorous, but that's a whole 'nother story...) The *Monotropa* absorbs the sugars from the fungus by "fooling" the fungus into thinking it's forming a mycorrhizal relationship-- but in fact the *Monotropa* is really parasitizing the fungus!. Thus the radiolabeled carbohydrates pass from the tree to *Monotropa* via their common mycorrhizal partner, in what is termed a source-sink relationship. In other words, the sugars flow from where they are made to where they are being used. Thus this is a three-way relationship between a photosynthetic tree, a mycorrhizal fungus, and a parasitic plant! You can see why some other terms for these monotropes that benefit from this *ménage à trois* are myco-heterotrophic plants, mycoparasitic plants, or epiparasitic plants

The fungi involved are very diverse. According to Martín Bidartondo, who has done some excellent work in Tom Bruns' lab at the University of California at Berkeley, *Monotropa uniflora* forms a relationship with *Russula* and *Lactarius* species. *Monotropa hypopithys* forms a relationship with various *Tricholoma* species. *Sarcodes sanguinea* (pictured below) forms a relationship with various *Rhizopogon* species. You can click here for Bidartondo and Bruns, 2001. "Extreme specificity in epiparasitic Monotropoideae (Ericaceae): widespread phylogenetic and geographical structure." This paper is very interesting and has a list of the species involved in all the relationships, along with triple phylogenies of all the symbionts. Very well done!

Phoradendron spp.: Mistletoe

Viscum alba, one of several species of plants with the common name of mistletoe. Photograph from http://commons.wikimedia.org/wiki/File:Mistletoe_P1210829.jpg, 2007 David Monniaux, available from GNU Free Documentation License.

From: http://www.apsnet.org/online/feature/mistletoe/

Introduction

Not many generations ago, before the advent of television and home entertainment centers, neighbors and relatives frequently visited each other for fellowship and did so especially during the holidays. A common custom at Christmas-time was for the homemaker to place a sprig of mistletoe above a door frame or hang it from the ceiling of the dwelling. During the frequent get-togethers, any female who lingered there was fair game for a harmless kiss from nearby males (Fig. 1). During the Yule season, mistletoe plants were sold in the market place and were as plentiful there as holly and the other widely used Christmas greenery. Today, greenery is still much used, but the use of mistletoe is seldom practiced even though almost everyone has heard of the custom of kissing under the mistletoe. In an era of televised and widely accepted sexual freedom such a custom seems sweetly quaint and naive and perhaps is not sophisticated enough to survive our modern moral standards.

Mistletoes are flowering plants (angiosperms) that obtain their nutrition by living on and parasitizing other plants. This relationship was observed across the European continent by ancient peoples who were so impressed with these plants that the mistletoe became interwoven into legends, myths, and religious beliefs (Fig. 2). It will be my purpose to acquaint the reader with the historic origins of some of these customs, especially with those on the European continent, and why they have survived to some extent to the present day.

What Are Mistletoes?

Mistletoes are parasitic plants that directly derive all or most of their nutrition from other flowering plants during most or all of their life cycle. There are approximately 3,000 parasitic angiosperms in 15 plant families, and almost all are dicotyledonous. Although many parasitic plants contain functional chlorophyll, they depend on their plant hosts for most, or at least some, of their carbon requirements and for all of their other nutrient and water needs. By parasitizing other higher plants, they have a competitive advantage over many other forms of life because they do not have to compete in soil for their water and nutrient needs. The mistletoes originated in the tropics, where soils are typically poor in nutrition and competition within the soil between plants and microorganisms is fierce. At the end of the last Pleistocene glaciation event of 18,000 years ago, there was an active northward and southward migration and evolution of some of the mistletoes. A general description of mistletoes may be found in Tainter and Baker (9). A good over-view of the evolution of the mistletoes is given by Kuijt (8). Information on parasitic plants can be found on the website of The International Parasitic Plant Society, as well as at the delightful "The Parasitic Plant Connection" website by Dan Nickrent.

The group of parasitic plants collectively known as the mistletoes are contained in four families, but only two of these, the *Viscaceae* and *Loranthaceae*, are of widespread importance. The family *Loranthaceae* is large and contains at least nine genera, most of which are abundant in the tropics. Most species have large, showy flowers and attack a variety of tree hosts. The family *Viscaceae* contains several genera, but only *Phoradendron* and *Viscum* are important in the legends and myths relating to the mistletoes. On the European continent *Viscum album* was the major species with which primitive man interacted and which formed the basis for many myths, legends, and religious beliefs. This may have been because the mistletoe plant grew on oak trees, which were revered by many early European tribes, or because it retained its leaves in autumn when the oaks defoliated. When fresh, its leaves had a yellowish-green color, and its stems were a yellowish color. After it was cut and dried, the plant developed a golden yellow hue. In the southern half of the U. S., where the European mistletoe does not occur, *Phoradendron* spp. are the common leafy mistletoe (Fig. 3). They are very similar in appearance to the European mistletoe.

Species of *Arceuthobium* are known as dwarf mistletoes because of their lack of leaves and reduced visible growth habit (Fig. 4). They were of much less importance in the mistletoe legends, partially because they were relatively inconspicuous, and also because they were not present to any great extent on the European continent. They were abundant on the North American continent, but were of only minor importance in folklore as Europeans settled in northeastern North America. With the native American people, mistletoe seems to have been more important for its pharmaceutical properties than for its role in folklore (6).

History

As ancient European people interacted with their environment and began to reason why certain things were the way they were, they developed an intense interest in trees. Possibly because of the many amenities derived from trees, and especially the oaks, trees came to be worshiped by these early Europeans. This eventually led to another Christmas ritual that has survived almost to the present, along with the seasonal ritual use of mistletoe. As part of a series of rituals, they burned logs around the time of the winter solstice (5). After conversion of the people to Christianity, the tradition of burning logs was changed to begin early on Christmas eve. A log was to be kept burning all night, and this culminated in a great celebration on Christmas morning. This yuletide custom of burning the Yule log was widely practiced until only a few decades ago, and probably ceased with the advent of centralized heating. The decline in the use of mistletoe probably was due to other factors.

Although the Greek philosopher Theophrastus (370 to 270 BC) described the common European leafy mistletoe, it was Pliny the Elder (23 to 79 BC) who wrote detailed descriptions of the attitude of some people toward the mistletoe. He recorded the widely held belief that whatever grew on the sacred oak was sent from heaven and, since mistletoe was only occasionally found on the oak, it was indeed cause for celebration when it was encountered (4). Pliny also recorded the belief that the mistletoe in winter contained the life of the oak after it had lost its leaves the preceding autumn. It was believed that the mistletoe plant was protected in some mystical sense from injury or harm. If it was cut from the oak, it retained some of these mystical powers, which could be channeled as healing powers. However, if it touched the ground after it was harvested, its healing powers would be lost.

While mistletoe played an important part in some later Greek and Nordic myths and legends, its relationship with the ancient Celts, who lived in ancient Gaul, Britain, and Ireland, is one of the earliest known examples of the importance of mistletoe. The Druids, who were the priests of a Celtic religious order, regarded the leafy mistletoes as having mystical properties and worshiped them (10). This belief in mystical properties was due, at least in part, to the fact that mistletoes often grew on the branches of the revered oak tree. In the autumn, as the length of the day decreased, religious significance was focused on the winter solstice, the shortest day of the entire year. The people observed that the mistletoe plants growing on the oaks retained their leaves while at the same time the oaks lost theirs. During the winter, the golden boughs of the mistletoe plant, with its yellow-green leaves and large white berries, seemed to be a remarkable phenomenon, and thus, the mistletoe plant was believed to have mystical properties.

The Druids also had a ceremony at Midsummer Eve which involved cutting a mistletoe plant from an oak tree with a golden sickle to initiate a ceremony in which animals and human beings were slain and burned (Fig. 5). When the Celts were later Christianized, they may have found it difficult to completely abandon their respect for the mistletoe plant, and it somehow became incorporated into a supposedly harmless custom which the early Christian church overlooked, even though it was widely practiced by its members.

While a feeling of veneration for mistletoe was widely shared by early European peoples, it was the Greeks who incorporated mistletoe into some of their myths and legends. The "Golden Bough" of Virgil's hero, Aeneas, was in fact mistletoe (5). Aeneas was arbitrarily chosen by the Latin poets to be the mythical progenitor of the Roman people. It was Aeneas' wish to visit hell, but on his way there he first had to pass through a vast and gloomy forest. Two doves guided him to a tree bearing a mistletoe plant (Fig. 6). He took the golden bough, and with its flickering light he was able to pass through the forest. When he emerged from the forest and showed the bough to the reluctant ferryman at the river Styx, both were immediately transported to the nether world. Such was the power of the mistletoe plant!

Another popular myth that involved mistletoe was that of the Norse god Balder (1). The myth held that the heavenly bodies, which included the gods, were created fresh every day. Odin, Balder's father, tried to help prolong Balder's life beyond that day and extracted a promise from all living beings not to harm him. However, he overlooked the mistletoe, and during archery practice, a rival gave an arrow made from a twig of mistletoe to Balder=s blind brother who accidentally shot Balder and killed him. This doesn't make too much sense to us today, but it probably made good logic at that time within the constraints of a myth. Probably as a result of trial and error, mistletoe plants were also found to have certain medicinal properties, and knowledge of these characteristics undoubtedly contributed to the mystical nature of mistletoe. In the ancient language of the Druids, mistletoe meant "all-healing." Some attributes were undoubtedly based in truth. However, others were certainly based on faulty reasoning. For example as late as 1900, an interesting use of mistletoe was for the treatment of epilepsy (10), which two millennia earlier was documented by Pliny. The rationale was that since mistletoe was rooted in the branch of a tree, and could not possibly fall to the ground, so too, an epileptic who took a decoction of mistletoe or carried it in his pocket could not possibly fall to the ground. A good review of pharmaceutical and other uses of mistletoes is given in Gill and Hawksworth (6).

From the Middle Ages to the last century, the literature is filled with examples of different uses for mistletoe plants, especially among rural people (4,5,10). It was cut, tied in bunches, and hung in front of cottages to scare away passing demons. It was hung over doors of stables to protect horses and cattle against witchcraft. In Sweden, it was kept in houses to prevent fire. It Italy it was believed to be able to extinguish fire. It was widely held to be a universal healer. As a potion it would make barren animals conceive. Even Pliny had known it was a cure for epilepsy, and that it could be used to promote conception. It healed ulcers if chewed. In Wales it was thought that, if placed under a pillow, mistletoe would induce dreams of omen. There were various customs in several countries that utilized mistletoe plants in rituals to find treasure. Collectively, these customs prove that mistletoe had a profound effect on people's lives and imaginations.

When Christianity became widespread in Europe after the 3rd century AD, the religious or mystical respect for the mistletoe plant was integrated to an extent into the new religion. In some way that is not presently understood, this may have led to the widespread custom of kissing under the mistletoe plant during the Christmas season, possibly relating to the belief in the effects on fertility and conception. The earliest documented case of kissing under the mistletoe dates from 16th century England, a custom that was apparently very popular at that time (10).

Much less is known about early historical aspects of the dwarf mistletoes. Because the plants of the dwarf mistletoes were small in size and not very abundant on the European continent, there is no record that ancient peoples of this region took any interest in the dwarf mistletoes. Although there are several species of dwarf mistletoes (*Arceuthobium*) in the European, Asian, and African continents, their most extensive development occurred in Central and North America, where there are over 3 dozen species of dwarf mistletoes (7). One of the smallest and most evolutionarily advanced species is *Arceuthobium pusillum*, which is found mainly in the spruce forests of eastern Canada and the eastern United States. The leafy mistletoes do not occur there because it is too cold in winter. However, even though the leafy mistletoes were not found there, Fernald (2) states that the women of the St. John and St. Lawrence River valleys wore sprigs of dwarf mistletoe in their hair while attending dances, following the European custom of women wearing leafy mistletoe in their hair, long before this species of dwarf mistletoe was known to science.

Biology of Mistletoe

Most parasitic higher plants use a similar process of infection. Upon germination, a root-like structure, called a radicle, emerges from the germinated seed and grows along the branch surface by a process known as thigmotropism. When it encounters an irregularity in the bark, the radicle will produce a swelling called a holdfast. A cementing substance may be secreted to bind the holdfast to the bark. A wedge-shaped structure, called a penetration peg, then forms to penetrate into the cortex of the host. Once established in the host's cortex, an intimate connection forms between phloem and xylem cells of the mistletoe and the phloem and xylem cells of the host, and the mistletoe plant then absorbs nutrients and water. These connections form structures called sinkers (Fig. 7).

One year or more after the mistletoe plant has infected the host, it will begin to produce the foliar parts of the plant. The mistletoe plant grows larger, producing a branched, golden-colored woody stem and yellow-green to dark green leathery leaves. The chlorophyll of leafy mistletoes is functional and photosynthesis is sufficient to supply all of their carbon needs. In mid-autumn small, round, pearl-like berries form (Fig. 8), and these enlarge to maturity in early winter. After they are mature, birds eat the berries and the seeds are carried away to begin new infections or the berries simply break off the plant and fall to lower branches to initiate new infections on the same tree or on understory vegetation.

Leafy mistletoe plants are perennial and remain alive within their respective host until the tree host, or the branch upon which it is established, dies. Since the leafy mistletoes can photosynthesize enough carbon to meet their needs, they only need to extract water, and whatever mineral nutrients the water contains, from their hosts. Thus, while they are obligate parasites in that they can only live and reproduce on a living host, they do not necessarily cause a debilitating nutritional drain on the host. A single infection on one branch, or only a few infections on an otherwise vigorously growing tree, seems to cause no noticeable harm to the tree. Often, though, that portion of a branch beyond the point of a single infection may become stunted in growth and even die prematurely. Multiple infections, sometimes dozens or even hundreds on a single tree, may produce a significant stress to the host tree that can either kill it outright or create stressful conditions attractive to secondary disease pests and insects that then cause premature death.

Seed dissemination of the leafy mistletoes is largely passive. The leafy mistletoes have a single-seeded berry which, when mature, contains viscin, a watery-sticky substance. Some local dissemination results when mature seeds are washed downward onto lower branches. Most distant dissemination occurs when birds feed on mistletoe berries or seeds. The seeds then either pass uninjured through the bird's digestive system or they adhere to the bird's plumage and feet and are removed during preening. Thus, dispersal of the seeds occurs wherever the bird defecates or preens its foliage.

The vsicin coating of each seed consists of numerous slender, spring-like cells that are embedded in a mucilaginous substance. Following dispersal, these viscin cells wrap around irregularities on the bark and help to glue the seed in place until it germinates and initiates a new infection. In Europe viscin was long used for the manufacture of birdlime, a sticky substance used to trap birds (10).

Infection by leafy mistletoes produces a slight, spindle-shaped swelling on the host at the infection site. If one makes a cross section through the swelling, the mistletoe tissues (mostly sinkers) will be a bright yellow-green or green in color and are easily distinguished from host tissues. The dark yellow-green to green woody stems are rather coarse in appearance and with relatively smooth but somewhat wrinkled bark.

The leafy mistletoes are generally not considered a serious enough threat to warrant the need for control measures. Exceptions might include fruit orchards, intensively managed plantations, or historically valuable or visually important specimen trees. In North America, the range of leafy mistletoes does not extend north of a line drawn across the United States from approximately Oregon to New Jersey, as they are susceptible to freezing temperatures. There are about a dozen species of *Phoradendron* in the U.S. They occur mostly on hardwood tree species, but some also occur on juniper, cypress, and incense cedar. *Phoradendron flavescens* attacks pecans in Florida, citrus in

Texas, and walnuts and persimmons in California. *Phoradendron juniperinum libocedri* and *P. bolleanum pauciflorum* cause significant losses on incense cedar and white fir in California.

Today, leafy mistletoes are commonly encountered on many urban and forest hardwood tree species. In a well-documented case, the famous horticulturist, Luther Burbank, imported the common European mistletoe into California in the early 1900s, and it has subsequently spread into the surrounding landscape on 24 species including willow, alder, poplar, elm, mountain ash, crabapple, and pear. In its native Europe, this mistletoe also attacks apple, almond, cherry, pine, fir, and poplar in parks, orchards, forests, and plantations.

Physical removal by cutting out infections of leafy mistletoes is certainly warranted in some situations. Care must be taken to remove all of the infection in the branch, as any living mistletoe tissues that remain are capable of regenerating into whole plants. The cut branches do not have to be burned or destroyed because the mistletoe plants die quickly. Some experimentation with herbicides suggests that these might be of value in less intensively managed forest situations, but are probably not economically justified.

A quarantine of infected plant materials would not seem to be difficult, as the mistletoes require living hosts in order to survive for extended periods of time in forms other than as seeds. The eradication of mistletoes around nurseries should ensure mistletoe-free seedlings. On the other hand, during the last century growth of mistletoe was encouraged in the western United States so that the foliage could be used as a supplemental cattle feed during especially harsh winters and coincidentally to harvest and sell during the Christmas season. In some gardens and arboreta mistletoe is propagated as a valued horticultural oddity.

Today, most of the mystical aspects of the mistletoes are not celebrated. Certainly we have a more realistic understanding of their pharmaceutical and medicinal properties. However, I think that most of the interest in them is based on their unique parasitic nature, perhaps similar to that observed in ancient times. But, our modern interest is based on the physiology of this parasitic nature in many different host-parasite interactions and the complex series of events during their evolution which resulted in their formation and development. Let us hope, though, that the old custom of kissing under the mistletoe at Christmas time is not lost.

Electronic Resources for Further Information

See the **Author's List of Related Links**

Literature Cited

1. Cox, G. W. 1883. An Introduction to the Science of Comparative Mythology and Folklore. Kegan Paul, Trench, and Company, London. [Reprint, Singing Tree Press, Detroit, 1968.

2. Fernald, M. L. 1900. *Arceuthobium pusillum* in the St. John and St. Lawrence valleys. Rhodora 2:10-11.

3. Frazer, J. G. 1890. The Golden Bough: A Study in Comparative Religion. Macmillan, London. [Reprint, Crown Publications, Inc., Victoria, BC, Canada, 1981.

4. Frazer, J. G. 1922. The Golden Bough: A Study in Magic and Religion, a New Abridgement from the Second and Third Editions. Macmillan, New York, NY. [Reprint, Oxford University Press, London, 1994.]

5. Frazer, L. 1924. Leaves From the Golden Bough. Macmillan, New York.

6. Gill, L. S., and Hawksworth, F. G. 1961. The Mistletoes: A Literature Review. USDA Forest Service, Technical Bulletin Number 1242.

7. Hawksworth, F. G., and Wiens, D. 1996. Dwarf Mistletoes: Biology, Pathology, and Systematics. USDA Forest Service, Agriculture Handbook 709.

8. Kuijt, J. 1969. The Biology of Parasitic Flowering Plants. University of California Press, Berkeley.

9. Tainter, F. H., and Baker, F. A. 1996. Principles of Forest Pathology. John Wiley & Sons, Inc., New York.

10. Von Tubeuf, C. 1923. Monographie der Mistel. Olderbourg, Berlin.

American Phytopathological Society
3340 Pilot Knob Road
St. Paul, MN 55121-2097
e-mail: aps@scisoc.org

LABORATORY EXPERIMENTS AND TECHNIQUES

Studies in mutualism

I. Introduction

Mutualism is one of several potential interactions or relationships which may occur when representatives of two species interact or encounter each other. They can be summarized by the following chart:

+ = enhances survival of symbiont

0 = does not affect survival of symbiont

- = decreases survival of symbiont

		Symbiont 2		
		+	**0**	**-**
	+	**Mutualism,** which may be either **obligate** or **facultative,** e.g. clownfish and sea anemones		
Symbiont 1	**0**	**Commensalism,** e.g. barnacles on the skin of whales.		
	-	**Predation,** if symbiont 1, as prey, gets eaten by symbiont 2, as predator; **Parasitism,** if symbiont 2 lives in or on symbiont 1.		**Competition,** in which symbiont 1 and symbiont 2 require the same limited resource. Over time, character displacement or specialization reduces the severity of direct competition

A. Termite intestinal flagellates

Introduction

    Trichonympha spp. and *Pyrsonympha* spp. are two genera of symbiotic flagellates that live in the intestines of some termites. Although termites can bite off pieces of wood and swallow them, they are incapable of chemically digest the cellulose into monosaccharides because they cannot synthesize cellulose. This enzyme is produced by these and other flagellates in the termite gut, enabling the termite to survive. In fact, when termites are exposed to an environment with enriched oxygen, will die of starvation because the increased oxygen concentration kills off the flagellate population in the gut. These flagellate protozoa are found nowhere else *except* in the termite gut. Therefore, this is an example of **obligate** mutualism.

    The flagellates themselves possess mutualistic bacteria which adhere to the pellicle of the flagellate. These bacteria engage in a synchronized movement, thereby providing locomotion to the flagellate cell.

    Vertical transfer of the flagellates is by regurgitation of food to each other. The regurgitated food contains the active flagellates. Termites have to be reinfected with digestive flagellates after each molt, because their hind guts (where the flagellates live) get shed with the skin.

Procedure

    1. Place 2-3 drops of Invertebrate Ringer's on a clean slide;

    2. Pick up a living termite with forceps and place it in the Invertebrate Ringer's;

    3. Use sharp probes to tease apart and spread the intestinal contents into the Invertebrate Ringer's solution. (There is nothing delicate to this part of the procedure. Just smush their little bodies apart.);

    4. Cover with a cover slip, and scan the slide under 100x, and then look carefully under the 400x.

Images:

http://workforce.cup.edu/buckelew/Trichonympha_sp_400x_other_termi.htm

http://bioweb.uwlax.edu/zoolab/Table_of_Contents/Lab-

2b/Termite_Gut_Flagellates/termite_gut_flagellates.htm

http://www.stcsc.edu/ecology/TermSymb.htm

Please draw any flagellates that you find on your slide:

II. Green hydra
Introduction

Hydra are a familiar organism in general biology courses. They are found in pools of freshwater, and are predators of microinvertebrates which are captured by the cnidoblasts. They undergo both asexual reproduction by budding, and sexual reproduction by the formation of gonads in separate individuals.

Chlorohydra viridissima is an unusual freshwater hydroid because the cells of its gastroderm contain mutualistic algal cells. Experiments demonstrate that radioactive carbon dioxide generated by the hydra cells diffuses into the algal cells and is used in photosynthesis to produce organic compounds.

Theoretically, therefore, the algal cells provide supplemental nutrition which could accelerate population growth during times of adequate sunlight. In this experiment, we will divide a population of green hydra into two groups. Both groups will be provided with brine shrimp *ad libidum,* but one group will be maintained in dark conditions while the other will be exposed to light. We will count mature hydras and buds each week over a month's time to determine whether the presence of light will induce a accelerate population growth.

Materials and Methods:
  A. Maintenance of hydra stock
      Hydra are very sensitive to pollutants in water. As soon as they arrive from the biological supply company, they must be transferred from the plastic containers into glass trays, covered with at least 3 cm of spring water. Spring water must be decanted off and replaced twice weekly;

  B. Culturing of brine shrimp for food;
      1. Hydra must be fed twice weekly, so two batches of brine shrimp must be prepared each week;
      2. Follow the instructions in preparing a batch of brine shrimp. That will involve adding iodine-free salt to spring water, mixing thoroughly, and then adding ¼ teaspoon of brine shrimp eggs, and then turning on the aerator. Let stand 24 hours or until most eggs have hatched.

C. Feeding brine shrimp to hydra;

    1. When the brine shrimp have hatched, turn off the aerator and place a lamp at one side of the flask. Brine shrimp, being positively phototactic, will swim towards the light.;

    2. Draw up the brine shrimp with a syringe and pour the contents through a strainer;

    3. Wash the brine shrimp twice with spring water;

    4. Add the brine shrimp to the glass trays and allow the hydra to capture their food;

    5. An hour later, decant off the water in the glass tray and replace with fresh water

An alternative to using brine shrimp for food is to order *Daphnia* spp., a freshwater microcrustacean, and pour directly into glass trays.

D. Counting hydra population

    1. Divide the hydra into two groups. Maintain one in a relatively shaded or dark area of the lab, and maintain the other adjacent to a lamp. Place a 1-liter beaker between the lamp and the culture trays to prevent heating of the water by the incandescent bulb;

    2. Each week, count the number of mature, independent hydra and the number of attached, immature buds;

    3. At the end of the experiment, graph the population changes by week.

	Population size					
Week	Shaded			Exposed to light		
Initial	Mature	Buds	Total	Mature	Buds	Total
1						
2						
3						
4.						

Images:

http://www.imagequest3d.com/catalogue/freshwater/pages/u021_jpg.htm

http://www.fcps.k12.va.us/StratfordLandingES/Ecology/mpages/green_hydra.htm

http://www.bioimages.org.uk/HTML/T76276.HTM

http://www.northern.edu/natsource/INVERT1/Hydra1.htm

http://www.cbu.edu/~seisen/Cnidaria/index.htm

Faunistic Surveys: Parasites of *Lepomis macrochirus*, the bluegill sunfish.

Parasites of Bluegill Sunfish, *Lepomis macrochirus* and of physid snails

Guidelines for lab report:
Introduction

- Description of hosts, *Lepomis macrochirus*, i.e. bluegill sunfish, and physid snails
- Description of parasites, and life cycles
    - *Posthodiplostomum minimum*
    - *Proteocephalus ambloplitis*
    - *Neoechinorhynchus thecatus*
- Purpose of this study:
    - To document incidence and intensity of parasites mentioned above among bluegills
    - To determine whether there is a correlation between *P. minimum* load and fish length
    - To determine distribution pattern of parasites among bluegills, whether it is clumped, uniform, or random.

Materials and Method
- Description of sampling site
- Description of sampling method
- Description of dissection protocol (concentrating on liver)

Results
- In text part of the Results section:
    - Mention the incidence and intensity of each parasite sp., and then refer to appropriate figures;
    - Discuss the results of the correlation analysis between fish length and *P. minimum* load, and then refer to appropriate figures.

Discussion
- What does 100% infection mean?

Faunistic surveys involve the complete examination and documentation of parasite species found in a host population. Information acquired by these surveys are used for the following:

1) To determine the presence of significant reservoir hosts of human parasites;

2) To document the incidence and intensity of parasite infections among intermediate or definitive hosts;

3) to determine the cause of massive and sudden mortality among economically important organisms being intensively cultured;

4) To test theories of population interactions, and

5) To describe previously unreported species.

These surveys require a complete and thorough post mortem examination of all organs and surfaces, preferably of freshly-killed animals. Using freshly-killed organisms is advantageous since the parasites will still be active and thereby will reveal their presence, and because the parasites will not have had sufficient time to migrate away from their normal site of infection.

If live hosts are unavailable for dissection, then preserved hosts are preferred over frozen ones, because the crystallization of water will distort or damage internal structures.

Special techniques are required to harvest all parasites from different host species. For example, old toothbrushes are used to dislodge ectoparasites from birds or mammals.

Knowing the organ from which particular parasite was removed assists in identification. Therefore, when a complete examination is performed of internal organs for endoparasites, each organ is placed into a separate dish full of the appropriate Ringer's solution. The organ is then carefully opened and examined for the presence of parasites.

Most endoparasites are small, white, and lacking in pigments. In order to identify a parasite, it is necessary to prepare it in such a way that diagnostic features are conspicuous and testable against the statements in a taxonomic key.

Complete preparation of a specimen for identification involves the three steps of fixing, staining, and mounting parasite on a slide.

We will be dissecting bluegill sunfish (AKA bream), *Lepomis macrochirus* from a local pond, probably in Shelby Farms. The parasites you are likely to encounter include the following:

1. *Posthodiplostomum minimum* metacercariae

Posthodiplostomum minimum centrarchi is a strigeid trematode whose metacercariae live in the visceral organs of bluegill sunfish, *Lepomis macrochirus* (Figure 1). Field studies have shown that prevalence of *P. minimum* among centrarchids quickly approaches 100% and that metacercariae become established in visceral organs when fish are extremely young, appearing in fish as small as 15 mm (Fischer and Kelso, 1990).

Laboratory studies have shown that the infection process will cause violent fin-fanning, brushing against aquarium walls and extensive hemorrhaging (Bedinger and Meade, 1967), and it can be lethal (Hoffman, 1958).

Metacercariae become fully developed and resistant to the effects of pepsin between 26 and 44 days, and it is probable that worms become infective to the definitive host at this time (Hoffman, 1958). Spall and Summerfelt (1970) state that metacercariae become infective 21 to 30 days post-exposure.

Field studies have yielded conflicting results in determining whether harmful or lethal effects of *P. minimum* metacercariae observed in laboratory studies also occur in nature. Bluegills infected with large numbers of grubs show decreased growth rates (Smitherman, 1964) or reduced mass for a given length (i.e., condition) (Hugghins, 1959). In contrast, infrapopulation size does not correlate with condition among bluegills (Lewis and Nickum, 1964), and the severity of white grub infection has no impact on hematocrit, relative weight, or growth in largemouth bass (Grizzle and Goldsby, 1996). Spall and Summerfelt (1970) conclude that most serious debilitating effects, which may result in death of the host, occur in the first 3 weeks following cercarial penetration (Eisen, 1999).

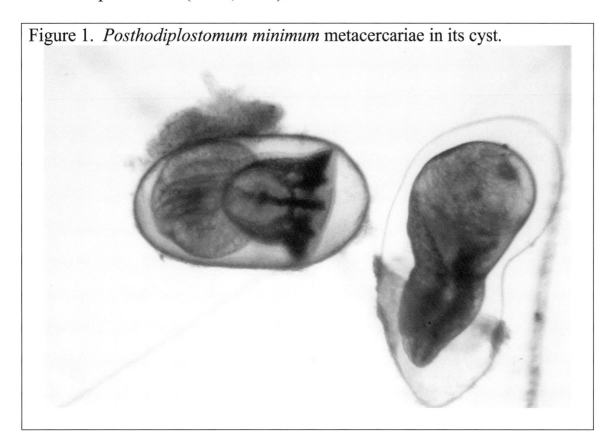

Figure 1. *Posthodiplostomum minimum* metacercariae in its cyst.

II.   *Neoechinorhynchus cylindratus* cystacanths

Neoechinorhynchus cylindratus ova are shed by largemouth bass, and are ingested by ostracods, a type of microcrustacean. The parasite larvae develop into an immature cystacanth stage, which will continue its development to a mature cystacanth stage in the bluegill sunfish (Bush, *et. al.*, 2001). The adult reaches sexual maturity when the infected bluegill is ingested by a largemouth bass.

III.   *Proteocephalus ambloplitis* pleroceroids

Largemouth bass is the definitive host for *Proteocephalus ambloplitis* as well. With this parasite species, the 1st intermediate hosts are copepods, another type of microcrustacean. The procercoid larvae undergoes partial development in the copepod, and it continues to the plerocercoid stage in small fishes, including the bluegill.

Other parasites which have been found among these bluegills prior to 2005 include *Spinitectus* spp. and *Camallanus* spp.

References Cited:

Eisen, S. (1999). "Distribution Patterns of *Posthodiplostomum minimum centrarchi* metacercariae among juvenile and mature bluegill sunfish, *Lepomis macrochirus*". Journal of the Tennessee Academy of Science 74(3-4):75-77.

PARASITOLOGICAL COLLECTION DATA

Name of host species: _____

Date of collection _____

Location of collection _____

Length of host (cm)_____   Weight of host (g)_____

Name of Parasite Species found in host	# of specimens found

Summary of Collection Data

Collection Date _____

Collection Site _____

Water temperature _____

Number of fish surveyed _____

Species	Incidence (% of hosts with that particular parasite)	Intensity (Average # of parasites per host)	Variance	Ratio of Variance/ Intensity	Correlation coefficient with fish length	Value of P
Posthodiplostomum minimum metacercariae						
Neoechinorhynchus thecatus larvae						
Proteocephalus ambloplitis larvae						
Spinitectus sp						
Camallanus sp.						

AN ADDENDUM TO FAUNISTIC SURVEYS: EXAMINING ROADKILL ANIMALS

Roadkilled animals sometimes harbor significant numbers of helminth and arthropod parasites. If you are interested in examining these types of animals, please note the following:

1) We are limited in the size of animal we can examine. A rule of thumb is to bring in only those animals which will fit into a grocery plastic bag;

2) Arthropod parasites, such as lice and fleas, migrate off the body surface when they detect a drop in body temperature. The best type of animal is a fresh roadkill which is still warm. Place the animal in the bag, spray it with a bug killer (if available), seal the bag by tying a knot and bring to the lab. When you bring the specimen, place it in a refrigerator;

3) Please do not bring in grossly mutilated specimens, e.g. if you can see the treadmarks of a tire on its squashed remains on the pavement, just leave it there;

4) You MUST dissect the animal within 24 hours of bringing it in. The specimen will become unsuitable to dissect after that time;

5) When you are finished dissecting the animal, please place the remains in a biohazard and drop the hazard bag into a wastebasket <u>OUTSIDE</u> the building.

"Ryegate, Montana artist Theodore Waddell, known for making art pieces from roadkill animals, put a coyote carcass on display this winter at the Cheney Cowles Museum in Spokane, Washington, Soon after, fly larvae hatched in the carcass and forced the museum to close until exterminators cleaned up." September, 1992. Source unknown, I am indebted to Robert Riccitelli for bringing this article to my attention.

6) UNDER NO CIRCUMSTANCES WILL WE ATTEND THE UPCOMING **ANNUAL HARVEST FESTIVAL AND ROADKILL COOK-OFF** HELD EACH SEPTEMBER IN MARLINTON MUNICIPAL PARK, POCOHANTAS COUNTY, WEST VIRGINIA, NO IFS, ANDS, OR BUTTS!!

THE PREPARATION OF PERMANENT SLIDES

I. FIXATION

The purpose of a fixative is to rapidly stop metabolic processed with a minimum of distortion or degeneration. As a result, fixatives consist of protoplasmic poisons. Three of the more common fixatives are the following:

AFA (Alcohol-Formol-Acetic acid)
formalin (formaldehyde USP)	10 parts
alcohol (95%)	25 parts
acetic acid, glacial	5 parts
glycerin	10 parts
distilled water	50 parts

Bouin's Picro-Formol-Acetic fixative
picric acid, sat. aq. sol.	75 parts
formalin (formaldehyde USP)	25 parts
acetic acid, glacial	5 parts

Schaudinn's Fixative
mercuric chloride, sat.aq.sol.	200 ml.
95% ethyl alcohol	100 ml.
acetic acid, glacial	12 ml.

Most specimens may be fixed by procedure 1. Thick specimens require procedure 2.

PROCEDURE 1. Place the cleaned and washed specimen into a Petri dish of saline. Place the Petri dish onto a hot plate and gently heat the dish until it is warm to the touch. The heating will relax the worm. When the worm has relaxed, pour off the saline and add the fixative. An alternative is to place the specimen into a dish with a small amount of cool saline, and then add several volumes of near-boiling fixative. Sudden exposure to hot fixative will also render most specimens fairly straight and extended.

PROCEDURE 2. Place the specimen on a slide with a drop of saline and wait until the specimen stretches out. Drop fixative on it, add a cover glass, and press down with sufficient pressure to flatten the specimen without crushing internal structures. Release the pressure and let the slide stand with enough fixative to float the cover glass until the worm is opaque. Then wash in a dish of fresh fixative.

Prior to dehydration and staining, the specimen must be washed thoroughly in water.

Once the specimens have been washed, they can be treated through each successive series of 30%, 50%, and 70% alcohol for 30 minutes each. Specimens may be stored in 70% alcohol indefinitely.

II. STAINING

Recipe for Erlich's Acid Hematoxylin:

Here is an alternative recipe for the above stain which is from Biological Techniques by Knudsen. If you have a chance to pick up a copy of this book it is a great reference, old but still very good!

Erlich's Acid Hematoxylin

Hematoxylin	2 gm
Absolute alcohol	100 ml
Distilled water	100 ml
Glycerin	100 ml
Acetic acid, glacial	25 ml
Potassium alum	10 gm

Dissolve hematoxylin in the alcohol and acid. Dissolve alum in heated water. Mix together. Place in stoppered bottle and age until it turns dark red (up to several weeks). Ready to use. Keeps for years.

The stains commonly used in parasitology are either alcoholic (e.g. Semichon's carmine) or aqueous (e.g. hematoxylin). Generally, staining is regressive, i.e. specimens are overstained and subsequently destined to the desire color. the skills of assessing degrees of staining and destaining are acquired though practice. For our purposes, hematoxylin will be used.

1. From 70% alcohol, return specimens to water by treating successively with 50% and 30% alcohol, 10 to 30 minutes each, depending on the size of the specimen.

2. Place in stain diluted with at least 9 parts of distilled water. Let stand for one hour.

3. Wash twice with water and treat successively with 30-50% alcohol, 10-30 minutes each.

4. Destain in 0.5 to 1.0% HCl in 70% alcohol until the specimens are light purple. Change the acid alcohol when necessary.

5. Wash in 70% alcohol, and then treat with 70% alkaline alcohol until the specimens turn a bluish color.

6. Replace with 80% alcohol and keep dish tight covered. Specimens may be left in 80% alcohol overnight.

III. DEHYDRATION, CLEARING, AND MOUNTING

1. Continue with the dehydration process by placing the specimen in each of the following baths for at least 30 minutes to several hours, depending on the thickness of the specimen:
95% alcohol
100% alcohol #1
100% alcohol #2
1 part 100% EtOH: 3 parts toluene
2 parts 100% EtOH: 2 parts toluene
3 parts 100% EtOH: 1 part toluene

2. Place a drop of the PERMOUNT mounting resin on the slide, and then place the specimen into the drop. Put a cover slip on top and apply light pressure spread the resin. Allow to dry.

3. Paste two labelling stickers on the slide. Write down the host

species and organ from which the specimen was removed on the left sticker. On the right sticker, write the species of the parasite, the date that the host was autopsied, and your initials. (You will need to use a taxonomic key to verify species names.)

IV.   Quantitative Analysis

Considerable research has been directed to determining the distribution patterns of parasitic organisms among their hosts.

Useful References

Eisen, S. 1999. Distribution Patterns of *Posthodiplostomum minimum centrarchi* metacercariae among juvenile and mature bluegill sunfish, *Lepomis macrochirus*. Journal of the Tennessee Acadmey of Science 74(3-4):76-77.

Margolis, L.; Esch, G.W.; Holmes, J.C.; Kuris, A.M.; Schad, G.A. (1982). The Use of Ecological Terms in Parasitology (Report of an *ad hoc* Committee of the American Society of Parasitologists. *Journal of Parasitology* 68(1):131-133.

SCHEMATIC DRAWING OF FIXING, STAINING, & MOUNTING
TIMES ARE APPROXIMATE

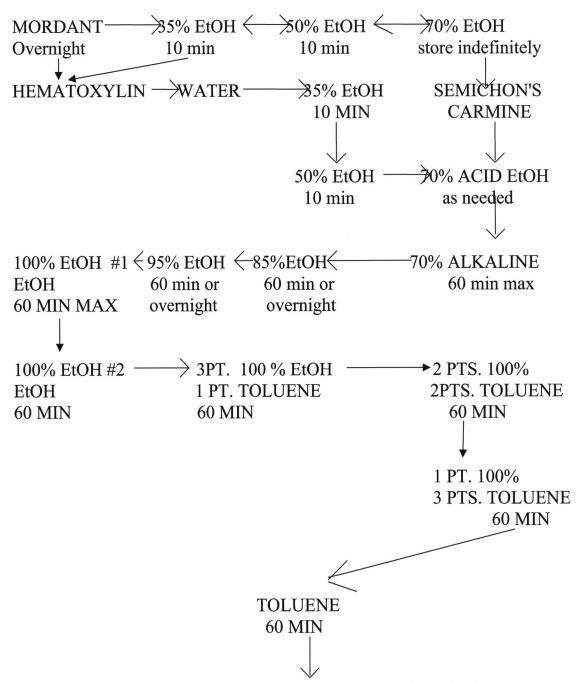

MORDANT ⟶ 35% EtOH ⟵ 50% EtOH ⟵ 70% EtOH
Overnight   10 min   10 min   store indefinitely

HEMATOXYLIN ⟶ WATER ⟶ 35% EtOH   SEMICHON'S
                                    10 MIN   CARMINE

                          50% EtOH ⟶ 70% ACID EtOH
                          10 min   as needed

100% EtOH #1 ⟵ 95% EtOH ⟵ 85% EtOH ⟵ 70% ALKALINE
EtOH   60 min or   60 min or   60 min max
60 MIN MAX   overnight   overnight

100% EtOH #2 ⟶ 3PT. 100 % EtOH ⟶ 2 PTS. 100%
EtOH   1 PT. TOLUENE   2PTS. TOLUENE
60 MIN   60 MIN   60 MIN

                                      1 PT. 100%
                                      3 PTS. TOLUENE
                                      60 MIN

          TOLUENE
          60 MIN

Place enough drops of Permount on slide, "bury" specimen in drop, cover with cover slip. Allow to dry overnight and then **LABEL**

Procedure for preparing permanent mounts of *Hymenolepis diminuta* ova and cysticercoids, *P. minimum* metacercariae, and *H. diminuta* adults.

Procedure for preparing permanent mount of *Hymenolepis diminuta* ova

1. Label your slide with adhesive label paper – *Hymenolepis diminuta* ova, fecal smear, your name, date;
2. **Gently** stir fecal material in jar;
3. Draw 1 drop of fluid onto slide;
4. Use probe to spread the drop no more than would be covered underneath a 22 x 22 mm cover slip;
5. Allow to air dry
6. Go to hood, and add 5-6 drops of Histoclad mounting medium
7. Gently place GLASS cover slip on mounting medium. You may gently press on cover slip with a pencil eraser to spread the mounting medium. Please avoid spreading the mounting medium past the dimensions of the cover slip.

Procedure for preparing a permanent mount of *Hymenolepis diminuta* cysticercoids or *Posthodiplostomum minimum* metacercariae

1. Label your slide with adhesive label paper – *Hymenolepis diminuta* cysticercoids, your name, date of collection, and species of animal from which the specimen it was extracted;.
2. Draw up cysticercoids with a dropper from the bottom of the vial with as little fluid as possible;
3. Verify with a microscope that you DO have at least one cysticercoid on the slide. If you do, allow to air dry;
4. Place in Coplin jar with haematoxylin stain for 60 seconds;
5. Place in Coplin jar with water for 60 seconds;
6. Allow to air dry,
8. Go to hood, and add 5-6 drops of Histoclad mounting medium
9. Gently place GLASS cover slip on mounting medium. You may gently press on cover slip with a pencil eraser to spread the mounting medium. Please avoid spreading the mounting medium past the dimensions of the cover slip.

Procedure for preparing a permanent mount of proglottids from

Hymenolepis diminuta **adults.**

1. Snip a small (10-15 mm) section or two from the worms in formaldehyde. Blot off as much fluid as possible;
2. Run the specimen through the stain and dehydrate;
3. Mount in Balsam cement

Procedure for preparing permanent slides of nematode, acanthocephalan, or arthropod specimens

The CMC and CMCP series of solutions are a convenient mordant/fixative/(stain)/mounting solutions for use with specimens which have a hard tegument, i.e. nematodes, acanthocephalans and arthropods.

1. Label your slide with adhesive label paper with the species name of the specimen, your name, date of collection, and species of animal from which the specimen it was extracted;
2. Place your specimen in the middle of the slide. If there is any solution associated with it (water, formalin, ethanol) remove as much of it as possible;
3. Use a glass rod as a "honey dipper" to place 5-6 drops of CMC on top of it, so that the specimen is completely covered. In order to allow the air bubbles to dissipate, leave the specimen *uncovered* for 24 hours;
4. Place a cover slip over the specimen, and let gravity and the viscosity of the mounting solution to draw the cover slip towards the specimen and to spread the mounting solution underneath the cover slip. Allow the specimen to dry for 24 hours;
5. Seal the edges of the cover slip with nail polish or lacquer.

Excystation *in vitro* of *Posthodiplostomum minimum* metacercariae

I. Introduction

Posthodiplostomum minimum is a digenean parasite which reaches sexual maturity in piscivorous birds, such as herons. Larvae transform into sexually mature, reproducing adults within 24 hours, and the infected host will be passing parasite ova within 24-48 hours after eating a fish with the metacercariae.

The first intermediate host for *P. minimum* is the freshwater snail, *Lymnaea* spp. A miracidium will penetrate the snail, and, through polyembryony, will produce cercariae which emerge from the snail in search of centrarchid fishes (sunfishes), especially members of the genera *Lepomis* spp. (e.g. *Lepomis macrochirus,* the bluegill sunfish, aka bream), *Micropterus* spp. (e.g. *Micropterus salmoides,* the largemouth bass), or *Pomoxis* spp., the crappie. Cercariae penetrate the skin of these fishes, burrow to visceral organs, and develop into a metacercarial stage, which remains dormant until the fish is ingested by a piscivorous bird.

In this exercise we will attempt to determine some of the host factors that cause excystation of *Posthodiplostomum minimum* metacercariae. In the process of excystation the juvenile worm is activated and frees itself from the tissue layers that surrounded it. Our working hypothesis is that bile salts, enzymes from the stomach and intestine, and temperature change may be factors involved in causing exycstation of *P. minimum* metacercariae. To test this hypothesis, we will inclubate cysticercoids in various solutions and compare the results with cysticercoids incubated in saline controls.

If *P. minimum* metacercariae are unavailable for this experiment, then *Hymenolepis diminuta* cysticercoids larvae collected from the hemolymph of beetles can be used (Figure 1).

Figure 1. *Hymenolepis diminuta* cysticercoid.

II. MATERIALS

Centrarchid fishes infected with the metacercariae (many will be found in the liver), or beetles infected with *Hymenolepis diminuta* cysticercoid larvae.

Deep-well depression slides

Several small Petri dishes with covers

Pasteur pipettes

Dissecting instruments and microscopes

Incubator adjusted to 38*C

Bile salt solution

Trypsin 1:250

Pepsin 1:10,000

Tyrode's or Earle's solution

Hydrochloric acid, concentrated

0.85% sodium chloride

Tyrode's solution

Component	Quantity
NaCl	8.00 g
KCl	0.20 g
$CaCl_2$	0.20 g
$MgCl_2$	0.10 g
$NaH_2PO_4 \cdot H_2O$	0.05 g

NaHCO$_3$ 1.00 g
Glucose   1.00 g

De-ionized water   1000 ml

Final pH should be 7.4

<u>Acid saline</u>
Component   *Quantity*
NaCl   8.50 g

Concentrated HCl 5 ml
De-ionized water 1000 ml

Final pH should be 1.5 - 2

<u>Invertebrate Ringer's</u>
Component   *Quantity*
NaCl   6.0 g
KCl   0.42 g
CaCl$_2$ 0.27 g
NaHCO$_3$ 0.20 g

De-ionized water   1000 ml

III.   <u>PROCEDURE</u>
    1. Dissolve 0.5 grams of bile salts and 0.5 grams of trypsin in 100 ml of Tyrode's solution, pH 7.4. (This step will have been done for your prior to lab.)
    2. Dissolve 1 gram of pepsin in 100 ml of 0.85% NaCl, add 0.5 ml of concentrated HCl; the final pH should be less than 2.0 (This step also will have been done for you prior to lab.)
    3. Preheat the solutions in small, covered dishes to 37*C for approximately 30 minutes.
    4. Dissect the beetles in depression slides with 1-2 drops of Invertebrate Ringer's solution. Remove the liver from the anesthetized fish and place it in a Stender dish with Invertebrate Ringer's. Tease the liver tissue apart and separate the metacercariae. Transfer them to a second Stender with clean Invertebrate Ringer's solution.
    5. Pipette 3-5 freed metacercariae with minimum of fluid into the

pepsin-saline solution and place in incubator at 37*C for 10 minutes. Pipette 3-5 metacercariae into 0.85% NaCl, and place in an incubator at 37*C for 10 minutes.   If we have an adequate number of metacercariae, then we can add two additional groups, those being a pepsin-saline group and a 0.85% saline group kept at room temperature.

6. Using the dissecting microscope and a Pasteur pipette, transfer half the metacercariae from the pepsin-saline solution into the Tyrode's-bile salts-trypsin solution. Transfer as little as possible of the pepsin-saline solution. Move the control group into some Tyrode's solution.

7. Put the dishes into the incubator at 37*C, and observe under the microscope every 5 minutes for 20 minutes until the worms have excysted.

Assume the following:
"0" = no discernible movement
"1" = hardly discernible movement
"2" = considerable movement
"3" = vigorous movement
"4" = continuous vigorous movement sufficient to break free of the cyst wall.

8. Determine the chemical conditions and environments required for the excystation of the metacercariae.

IV. Data

Degree of movement:	Treatment #1: Temp: Environment(s):	Treatment #2: Temp: Environment(s):	Treatment #3: Temp: Environment(s);
Baseline - immediately after setting up solutions			
After 10 minute initial incubation			
20 minutes post-removal			
30 minutes post-removal			
40 minutes post-removal			

EXTRACTION AND STAINING OF MONOCYSTIS SPP. FROM EARTHWORMS

Adapted from Sheridan, P. 1986. *Monocystis*: Earthworm Parasite. The American Biology Teacher 48(1):20-23.

Classification:

The family Monocystidae contains more than 183 species that occur worldwide in a variety of genera of earthworms. Monocysts is classified as an acephaline gregarine sporozoan; that is, its body is nonseptate, the mature trophozoites are large and extracellular, and it is a protozoan whose immature stage develops a resistant wall that encloses the sporozoites (Levine 1977). The genera of Monocystidae are identified partly by the shape and structure of the trophozoites and partly by the shape and arrangement of the spores. There are no intermediate hosts, and all members of the family are parasitic in invertebrates (Nobel 1974).

Life Cycle:

Monocystis is ingested by an earthworm as a spindle-shaped spore. In the gut of the earthworm the spore liberates eight, motile, sickle-shaped sporozoites, each of which may penetrate the gut wall, pass through the body to the testes and enter a clump of immature sperm cells. The sporozoite matures into the trophozoite by consuming the host protoplasm with the protein and nucleic acids seemingly necessary for later cell division. The worm's sperm cells disintegrate, leaving the filaments on the surface of the trophozoite and giving it the appearance of a ciliated organism. The mature trophs leave the tissue, make their way to the seminal vesicles, group in unfused mating pairs (syzygy), and surround themselves with a polysaccharide capsule as the gametocytes within a gametocyst. Each gametocyte undergoes multinucleation producing a number of gametes. The gametocyte membrane disintegrates and each of the gametes from one parent can then fuse with a gamete from the second parent, resulting in the formation of zygotes (sporoblasts). Within the cyst each zygote produces a thick wall around itself and becomes a spore, while within the spore nuclear division produces eight falciform sporozoites. Some spores escape to the soil via the vasa deferentia while other spores may be transferred into the seminal receptacles of mating earthworms, or if the cocoon is broken, may also pass into the soil to infect the adult earthworms which ingest them. Parasitized worms may also be eaten by birds and, with the spores passing through the digestive tract unchanged, the dissemination of Monocystis is assured.

Availability and Occurrence:

Monocystis is available at all times of the year. Research indicates a high degree of infestation in general (Levine 1977; Dukhlinska 1977), and more than one species can be found in a single host.

References:

Levine, N.D. (1977). Revision and checklist of the species of the aseptate gregarine family Monocystidae. Folia Parasitol., 24(1):1-24.

Noble, E. & Noble, G. (1974). Parasitology (3rd ed.). Philadelphia, PA.: Lea and Febiger.

Duklinska, D. (1977). On the distribution of gregarines in lumbricid earthworms from Bulgaria. Acta Zool. Bulg., (7):49-59.

Study Preparations:

To obtain specimens for study, the living earthworm is anesthetized in (6%) alcohol or chloroform. The anesthetized worm is pinned down in a dissecting pan and opened with a scalpel along the mid-dorsal line from the mouth to the clitellum. Peel back the body wall and pin it to the pan, exposing the internal organs. The mature seminal vesicles are the three pairs of prominent white sacs surrounding the digestive tract at the level of the 10-15th segments. The seminal receptacles found in the ninth and tenth segments will also provide parasites in large numbers.

Wet Mount Smear:

A quick temporary preparation is the wet mount smear. This should be performed first in order to determine the degree of infestation so that time is not wasted working with material which contains few parasites. Remove a seminal vesicle with forceps and smear the organ across the viewing area of the slide. Add a drop of 0.7 percent saline or stain, such as methylene blue, add a coverslip, and examine for stages in the life cycle. The small highly refractile spores are lemon-shaped and can be seen to contain a number of sporozoites. The larger trophozoites attain a length of 300 micrometers or more, readily absorb stain and look like comets or fuzzy tadpoles. The spherical cysts are easily found and the various types of cysts can be differentiated from one another by their contents. Young gametocysts will contain two discrete dense cells while in the mature gametocysts the contents are granular and the two parent cells are not easily distinguished from each other. The zygocysts can be recognized by the appearance of many zygotes, each of which shows a distinct nucleus, cytoplasm and cell membrane. The cysts containing the highly refractile spores are easily recognized and quite numerous.

Permanent Smears:

A permanent stained mount can be prepared with the following procedure:

1. Smear the seminal vesicle on the slide and air dry.
2. Stain in Harris's hematoxylin for about three minutes.
3. Rinse briefly in tap water.
4. Blue with 0.1% sodium bicarbonate for about one minute or until distinctly blue in color.
5. Rinse throughly in tap water.
6. Counterstain with eosin in 70% ethyl alcohol for about 45 seconds.
7. Rinse in 95% ethyl alcohol by dipping momentarily and then draining. Repeat a few times.
8. Dry (or dehydrate in alcohol and clear). Simple drying will suffice as the smear is thin and has been air-fixed.
9. Mount in a resinous medium (e.g. PERMOUNT) with a coverslip, label and allow to harden. (Hardening will take about 24 hours.)

Life Cycle of Mosquitos and Larvicidal Effects of Light Diesel (Golden Bear) Oil and Mosquito Bits™

I. Introduction

Mosquitos are important in their roles as parasitoids and vectors for disease. The list of diseases transmitted by mosquitos includes viruses (e.g. yellow fever), protozoa (*Plasmodium* spp.) and helminths (*Dirofilaria immitis*).

The life cycle includes 4 larval (instar) stages, and 1 pupal stage before the mosquito reaches sexual maturity in the imago stage. The duration of developmental stages is inversely proportional to ambient temperature. Although the larvae and pupae are aquatic, they breathe air through an air 'trumpet' or tube.

The fact that the larvae breathe air provides a means to reduce the populations of larvae in water. A common practice in mosquito larval control is the use of a light oil to block the penetration of the air tube through the air-water interface. The particular brand used by the Memphis Mosquito Control Board is called "Golden Bear" oil. The oil is dispersed via backpack sprayers over wetland areas. Although it evaporates within three hours of application, that time is sufficient to asphyxiate the larvae in the water.

There is also a new product, **Mosquito Bits™**, which is pelletized *Bacillus thuringensis*, a bacterium which secretes an endotoxin which induces internal bleeding.

A common measurement of bioassays is the LT_{50}, the time required for 50% of the test organisms to die within the experimental period. This study involves a bioassay to determine the efficacy and LT_{50} of Bear Oil and Mosquito Bits.

II. Materials and Methods

Prior to the beginning of the laboratory experiment, a fermented mixture of grass clippings and water was prepared and allowed to stand for a minimum of three days. This mixture was placed in to a brown dishpan to serve as bait for female mosquitos to lay their eggs. Over a weekend to be specified later, the dishpans will be returned to the lab, presumably with a large number of eggs in it. By the following Monday morning, larvae should have hatched in the dishpan and are ready for the experiment.

Remove the larvae from the dishpan with the "grass-clipping brew", and place equal numbers (10) into each of three Stender dishes, each half-filled with either aged tap water or pond water. Add 5 ml of Bear Oil to one, and add one pellet of Mosquito Bits to the third. The application rates are the following:

Therefore, each group should have 3 Stender dishes labeled as follows:
1. Control (plain water)
2. Treated with Bear Oil
3. Treated with Mosquito Bits

During class time, observe the behavior of the larvae in each of the dishes. Furthermore, observe all dishes once a day to determine the number of days required to reach the pupal stage and to reach the adult OR to determine how many of the larvae have died. Remove and count the dead larvae. Please note that we will not identify the species found in the dishes.

Record the cumulative number of dead larvae in each of the Stender dishes, and the number of days required for surviving larvae to reach the pupal and adult stages. Enter the data on the chart on the following page.

	Dish #1: Control	Dish #2: Treated with Bear Oil	Dish #3: Treated with Mosquito Bits
Initial number (Day 0)	10	10	10
Cumulative percentage of dead larvae at: Day 0	0	0	0
Day 1			
2 days			
3 days			
4 days			
5 days			
6 days			
7 days			
8 days			
# days to reach pupal stage			
# days to reach adult stage			

CONCENTRATION OF CYSTS, EGGS, AND LARVAE BY
CENTRIFUGATION-SEDIMENTATION AND FILTRATION
(from Strickland, G. Thomas, 1984. Hunter's *Tropical Medicine*,
edition 6, W.B. Saunders Company)

This technique is quoted verbatim from page 992.

Formalin-Ether Sedimentation. This is the method of choice for general use. It concentrates helminth larvae, eggs, and protozoan cysts.

Preparation. Prepare a 10% solution of formalin in water.

Procedure.
1. Partially comminute the entire fresh stool with an appropriate amount of wter or saline solution in the stool container. Add enough fluid to make it possible to recover 8 to 10 ml of strained emulsion, which when centrifuged, will yield about 0.5 to 0.75 ml of fecal sediment.
2. Strain the emulsion through 2 layers of moist gauze, and collect it in a 15-ml pointed centrifuge tube. A cone-shaped paper cup with the point cut off can be substituted for the glass funnel. The straining may be omitted to simplify the procedure.

(Short cuts may be taken that generally result in a satisfactory specimen fo examination. In such cases, substitute the following for steps 1 and 2 above: [1a] Emulsify approximately 2 ml of feces in 20 ml of water. [2a] Strain the emulsion through 2 layers of moist gauze or a commercial plastic tea strainer into a 15-ml conical centrifuge tube.)

3. Centrifuge at 1500 rpm for 2 minutes, or at 2000 to 2500 rpm for 1 minute. Decant the supernatant fluid. The resulting sediment should be about 0.75 to 1 ml. Add water, shake, and recentrifuge. Repeat this process until the supernatant fluid is clear.
4. Decant the supernatant fluid, and add 10 ml of the 10% formalin solution. Mix thoroughly with the aid of an appliator stick and *allow to stand for 5 minutes*.
5. Add 3 ml of acetone. Centrifuge at 1500 rpm for about 2 minutes, and allow the centrifuge to stop smoothly and slowly. Four layers should be seen in the tube: a small amount of sediment, which contains most of the parasites; a layer of formalin; a plug of fecal debris on top of the formalin layer; and a layer of ether at the top.
6. Loosen the plug of debris from the sides of the rube by ringing it with an applicator stick, and then carefully decant the top 3 layers.

Swabbing the sides of the tube with a cotton-tipped applicator stick gives a cleaner preparation for examination.

7. Mix the sediment with the small amount of fluid that drains back from the sides of the tube. Remove the deposit with a pipette, and place it on a microscope slide. Cover with a coverglass, and examine. The preparation may be stained with Lugol's iodine solution to identify any cysts present.

The formalin-ether sedimentation technique is excellent for the detection and identification of protozoan cysts, helminth eggs, and larvae of almost all intestinal worms. The technique is also very useful for examining stools containing fatty substances that interfere with the performance of the zinc sulfate centrifugal flotation method. It is however, *not* satisfactory for trophozoites.

Formalin-ether with Preserved Specimens. Formalinized specimens can also be treated by the method just described, but tap water may be substituted for the formalin solution. After the final centrifugation, both unstained and iodine-stained smears can be prepared for microscopic examination in the usual way.

Helminth parasites of hydrobiid snails
To be conducted at the Gulf Coast Research Lab

Introduction:

    Hydrobiid snails are commonly found in soft substrate tidal marshes (Figure 1). They graze on diatoms and other microalgae found on sea grasses, e.g. *Spartina* spp. Despite their small size, they serve as first intermediate hosts for a variety of digenean parasites, including *Phagicola diminuta* and *Microphallus turgidus*.

    In a survey of hydrobiids of tidal marshes along the Skidaway River in Georgia, Pung (*et. al.,* 2008) found 4 types of larval trematodes:
1) An ocular monostome cercariae, 2.2% overall prevalence;
2) Type I xiphidiocercariae, 1.3% overall prevalence;
3) Type II xiphidiocercariae, 2.2% overall prevalence, probably corresponding to *Microphallus turgidus*;
4) A sanguinicolid cercaria, only 1 specimen observed among 4201 *Spurwinkia salsa* and 139 *Onobops jacksoni* sampled.

An unusual feature of the pattern of parasitism is that each hydrobiid will be infected with only one species of digenean parasite.

Digenean trematodes exert considerable deleterious effects on their snail hosts. For example:
1) Alda (*et. al.,* 2010) have found that infection of the South American hydrobiid *Heleobia australis* by *Microphallus simillimus* leads to a narrowing of shell width;
2) Lauckner (2005) has found that the digeneans infecting the periwinkle *Littorina littororea* on the German North Sea coast can cause complete parasitic castration, reduction in longevity, and size-related differential mortality among snails in different age classes;
3) Probst and Kube (1999) found evidence of gigantism among infected *Hydrobia ventrosa*. They interpret this phenomenon as a side effect caused by parasitic castration induced by the 7 taxa of digenean trematodes infecting this species.

The purpose of this study is to three-fold:
1) To determine the species of hydrobiid snails found in Simmons Marsh, Ocean Springs, MS;
2) To determine the prevalence and distribution of digenean parasites among these hydrobiid snails;
3) To compare the shell length and width of infected and uninfected snails.

Materials and Methods

Snails will be collected with a hand net and brought back live to the lab. The species of each individual snail will be determined according to Heard, *et. al.* (2002), and its shell length and width will be recorded.

Digenean species will be identified according to Heard and Overstreet (1983). ANOVA tests will be employed to compare the mean shell lengths and shell widths of uninfected snails with snails infected with each of the parasite species.

Specimen #	Shell Length (mm)	Shell width at widest point (mm)	Parasite species. (Write down N.A. if no parasites were found.)
1.			
2.			
3.			
4.			
5.			
6.			
7.			
8.			
9.			
10.			
11.			
12.			
13.			
14.			
15.			
16.			
17.			
18.			
19.			
20.			

References Cited:

Alda, P.; Bonel, N.; Cazzaniga, N.J.; Martorelli, S.R. (2010). Effects of parasitism and environment on shell size of the South American intertidal mud snail *Heleobea australis* (Gastropoda). Estuarine, Coastal and Shelf Science 87:305-310.

Heard, R.W.; Overstreet, R.M. (1983). Taxonomy and life histories of two North American species of *"Carneophallus"*, (= *Microphallus*) (Digeanea: Microphallidae). Proceedings of the Helminthological Society of Washington 50:170-174.

Heard, R.W.; Overstreet, R.; Foster, J.M. (2002). Hydrobiid snails (Mollusca: Gastropoda: Rissooidea) from St. Andrew Bay, Florida. Gulf and Caribbean Research 14:13-34.

Lauckner, G. (1984). Impact of trematode parasitism on the fauna of a North Sea tidal flat. Helgoland Marine Research 37(1-4):185-199.

Pung, O.J.; Grinstead, C.B.; Kersten, K.; Edenfield, C.L. (2008). Spatial Distribution of Hydrobiid Snails in Salt Marsh along the Skidaway River in Southeastern Georgia with Notes on Their Larval Trematodes. Southeastern Naturalist 7(4):717-728.

Probst, S. & Kube, J. (1999). Histopathological effects of larval trematode infections in mudsnails and their impact on host growth: what causes gigantism in *Hydrobia ventrosa* (Gastropoda: Prosobranchia)? Journal of Experimental Marine Biology and Ecology 238(1): 49-68.

Analysis of dog fecal samples collected from dog parks in the Memphis area

Introduction

Despite the diligence of companion pet owners, a significant percentage of dogs can accumulate helminth infections. For example, in a study conducted by Ciordia and Jones (1956), 68% of dogs in the Knoxville, TN area were infected with *Dipylidium caninum* and 16% were infected with *Taenia pisiformis*. The number of parasites found in each animal ranged from 1 to 178 and 1 to 18, respectively. In a study of fecal specimens from 4,058 dogs admitted to the Louisiana State University Veterinary Teaching Hospital and Clinics, Hoskins and Malone (1982), found that one of more species of parasites was identified in over 50% of dogs. The most frequently encountered parasites were hookworms, found in 38.5%, and whipworms found in 14.9%.

It may be possible to discern differences in the percentage of infected dogs from fecal samples collected from a variety of dog parks. Overton Bark, for example, is a relatively small dog park designed for residents of Midtown Memphis and the surrounding area. It has an enclosed area of 1.3 acres, divided into two sections for large and small dogs. It is located in the 38104 zip code, where the median income is $30,256. Shelby Farms is located in East Memphis, and has an area of 4,500 acres, including lakes, natural forests and wetlands. It is considered one of the largest urban parks in the United States, and is located in the 38134 zip code, where the median income is $47,341. Johnson Park is located in Collierville, TN, occupies 278 acres, and has natural areas, wetlands, and three lakes. It is in the 38107 zip code, which has a median income of $79,259.

Assuming that companion pet owners will not drive too far to take their dog for a walk, perhaps we can hypothesize that the prevalence of parasitic infections among dogs will be inversely proportional to the median income of their owners, as reflected by their median incomes. That means that dogs owned by families in Johnson Park area will have a lower prevalence of parasitic infections than the dogs owned by families in the Overton Bark area.

We can test this hypothesis by collecting fecal samples at each of these sites and analyzing them for helminth ova.

Procedure

- Half-fill each specimen bottle with 4% formaldehyde;
- Place about 1 tablespoon of sample into each bottle. Label each bottle with a distinctive number;
- Fecal samples will be examined via a direct fecal smear. Vortex each bottle for 15 seconds with a vortexer. (Make sure the lid is secure!) Place 3 drops of the solution onto a microscope slide, cover with a cover slip, and examine the slide with 100x magnification. If anything looks like an ovum, verify by using 400x magnification:

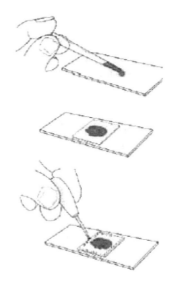

- Collect information from each student to compile a comprehensive list of analyzed fecal samples;
- Calculate the prevalence of each parasite, and a frequency histogram of parasitic infections.

Data sheet for fecal samples (Duplicate this data sheet for each sampling location)

Sample location: _____

Sample date: _____

Sampling location	Sample No.	Hookworm	*Trichuris vulpes*	*Toxocara canis*	*Strongyloides*	Tapeworm	Other-please specify	Total Number parasite species
	1							
	2							
	3							
	4							
	5							
	6							
	7							
	8							
	9							
	10							
	11							
	12							
	13							
	14							
	15							
	16							
	17							
	18							
	19							
	20							
	21							
	22							
	23							
	24							
	25							
	26							
	27							
	28							
	29							
	30							

References Cited

Ciordia, H.; Jones, A.W. (1956). The incidence of intestinal helminths in dogs and cats in Knoxville, Tennessee. Journal of the American Veterinary Medical Association 128(3):139.

Hoskins, J.D.; Malone, J.B. (1982). Prevalence of parasitism diagnosed by fecal examination in Louisiana dogs. American journal of veterinary research 43(6):1106-1109.

Insects which induce gall formation in plants

Introduction

From: http://en.wikipedia.org/wiki/Gall

Galls or *cecidia* are outgrowths on the surface of lifeforms. Plant galls are abnormal outgrowths[1]of plant tissues and can be caused by various parasites, from fungi and bacteria, to insects andmites. Plant galls are often highly organized structures and because of this the cause of the gall can often be determined without the actual agent being identified. This applies particularly to some insect and mite plant galls. In pathology, a gall is a raised sore on the skin, usually caused by chafing or rubbing.[2]

Insect galls are the highly distinctive plant structures formed by some herbivorous insects as their own microhabitats. They are plant tissue which is controlled by the insect. Galls act as both the habitat and food source for the maker of the gall. The interior of a gall can contain edible nutritious starch and other tissues. Some galls act as "physiologic sinks", concentrating resources in the gall from the surrounding plant parts.[3] Galls may also provide the insect with physical protection from predators.[4][5]

Insect galls are usually induced by chemicals injected by the larvae or the adults of the insects into the plants, and possibly mechanical damage. After the galls are formed, the larvae develop inside until fully grown, when they leave. In order to form galls, the insects must seize the time when plant cell division occurs quickly: the growing season, usually spring in temperate climates, but which is extended in the tropics.

The meristems, where plant cell division occurs, are the usual sites of galls, though insect galls can be found on other parts of the plant, such as the leaves, stalks, branches, buds, roots, and even flowers and fruits. Gall-inducing insects are usually species-specific and sometimes tissue-specific on the plants they gall.

Gall-inducing insects include gall wasps, gall midges, gall flies, aphids (such as *Melaphis chinensis* and *Pemphigus betae*), and psyllids.

Hackberries are subject to nipple galls because of the larvae of several

244 | Page

species of Psyllidae, insects of the Order Hemiptera (Figure 1).

Figure 1. Psyllid larva extracted from a hackberry nipple gall

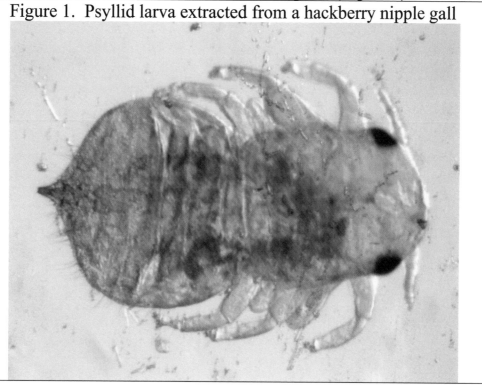

The adults are called jumping plant lice because the resemble miniature cicadas. In the spring, adults emerge from the leaf litter to mate and deposit eggs on newly-formed leaves. After hatching, the nymphs feed on leaves and produce the distinctive gall, which average 4 mm in width and 6 mm in height, on the underside of leaves. The nymphs will remain inside the leaves through the summer, and will emerge in September. The species overwinters as adults within bark crevices.

Procedure
- Collect branches from hackberry trees, which are located on the west side of Stritch Hall, and bring them back to the lab.
- Remove a leaf, and then use scissors to cut as much away from an individual gall as possible;
- Place the gall in a Petri dish half-filled with water. Use a scalpel to gradually cut away the leaf tissue, and use needle-nosed forceps to remove the nymph;
- Prepare a permanent slide of the nymph by placing it on a microscope slide;

- Use a glass rod as you would a honeydipper to apply 4 drops of CMC-10 over the nymph;
- Gently place a glass coverslip over the CMCP-10, and allow gravity to spread the CMCP-10;
- Label your slide.

References:

Buchanan, C.K. and Heard, S.B. (1998). Larval performance and association within and between two species of hackberry nipple gall insects, Pachypsylla spp. (Homoptera: Psyllidae). The American Midland Naturalist 140(2): 351.